KW-222-876

HOME
SECURITY
AND
PROTECTION

HOME
SECURITY
AND
PROTECTION

Robert Traini

WILLOW BOOKS
Collins
8 Grafton Street, London
1984

Willow Books
William Collins Sons & Co Ltd
London · Glasgow · Sydney
Auckland · Toronto · Johannesburg

First published in Great Britain 1984
© Robert Traini 1984
All rights reserved. No part of this publication may be
reproduced or transmitted in any form or by any means,
electronic or mechanical, including photocopying, re-
cording or any information storage and retrieval system
now known or to be invented without permission in
writing from the publisher.

Cover illustration by Ian Stephen (Studio Briggs Ltd)

Traini, Robert
Home security and protection.
1. Dwellings – Security measures
I. Title
643'.16   TA9745.D85

ISBN 0 00 218093 6

Filmset in Imprint by Butler & Tanner Ltd, Frome and London
Printed in Great Britain by William Collins Sons & Co Ltd, Glasgow

Contents

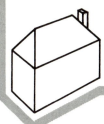

Locking devices (*left to right, top to bottom*): mortice deadlocks; door closer, rim latches; devices forming electrical mechanical locking systems; various padlocks; safe deposit box lock, two furniture locks, keys, two microswitch locks, cam lock, shunt lock.

1 What does security mean?

Security covers a very wide field in which the risks and the threats vary greatly. This book is concerned only with the hazards presented in a domestic situation and the practical and economical means available to protect the home, its contents and its occupants from the effects of criminal attack and fire. It must never be forgotten that these two aspects go hand in hand: for although it is the object of security to keep the potential intruder out, this must somehow be made compatible with the equally vital need to ensure that, in the event of fire, the occupants are able to escape quickly. Recent domestic tragedies in which people have died when trapped inside their blazing homes have highlighted the importance of careful planning to ensure that effective protection from the villain does not create an even greater threat from fire and smoke. Quite apart from the blaze caused accidentally, arson sometimes accompanies both burglary and vandalism, or is perpetrated on its own.

It is only comparatively recently that there has been a scientific approach to security of the home and the family. The first generation of burglar alarm systems were merely adaptations of systems designed originally for the security of industrial and commercial premises in which the hazards were different and cost-effectiveness was on a quite different scale. Even now, many conventional alarm systems are still such adaptations. But more and more of them are now being created specifically for domestic situations. This welcome progress has itself brought problems for the home-owner. Some of these systems do not by any means live up to the claims which are made for them. Some of them are marketed at prices far beyond their worth. And, with some 200 or more companies now manufacturing alarm systems, the prospective buyer has little opportunity of comparison and assessment.

Another disquieting feature is the increase in the number of 'cowboy' operators out to make a 'fast buck' by cashing in on the growing demand for home protection devices and pressuring householders to buy installations they do not need and that do not perform the task for which they are designed. In its 1983 annual report, the Advertising Standards Authority criticised some manufacturers of home security equipment for overstepping the mark by being 'excessively emotive' in their advertising. The Authority expressed concern about the combination of unsolicited literature and copy that was blatantly aimed at frightening consumers into buying and fitting intruder alarms. In all, nine offenders were investigated by the Authority and the complaints against them were upheld in every case. Direct mail leaflets had been used in seven out of the nine cases.

Remember always that you do *not* have to live in a fortress in order to deter thieves. Before you consider the possibility of installing an alarm system, which may be very expensive if you employ professional installers, bear in mind that there is a basic level of security which must be in existence first and which will itself provide you with a fairly high degree of protection under normal circumstances.

Security is really an attitude of mind and the techniques now available to translate that attitude into physical protection can be effectively employed only if *all* the risks are recognised and each separate measure is carefully assessed and introduced to form part of a total security plan. That plan must be designed to meet the particular circumstances of your home.

There are three very important factors to take into consideration when planning security for your home. Many home-owners decide against the installation of an alarm system on the grounds of expense alone. But although it is an important factor, since it is inevitable that the value of the premises and the contents of the home are weighed against the cost of a protective system and few people would spend a lot to protect a little, this may be a superficial view. The insurance company will recompense the victim financially for the intrinsic value of any items which have been stolen and for the repair of damage to the home. The second factor is that nothing can compensate for the loss of, or damage to, those other cherished personal belongings which are irreplaceable and of great sentimental value: items that are part of the very fabric of family life. Items which have taken a lifetime to create and to collect may be stolen or destroyed in seconds by an intruder. And nothing can compensate for the violation of the privacy of the family home. Thirdly, the home-owner today must consider how much he is prepared to pay to ensure the peace of mind of not only himself, but of his wife and family when he is away from home. The same consideration is perhaps even more important for the single parent, which often means a woman alone with one or more young children. The rising tide of burglary alone is frightening, but the ugly features which often accompany it, from mindless vandalism to arson, or violence as vicious as rape or murder, call for something more than just basic security.

Much has been written and spoken recently about crime and the fear of crime. It has been shown statistically that the fear of crime is often very much greater than the actuality. Statistics are cold and unsympathetic, anyway, but some amazing things can be done with them. The British Crime Survey, based on the questioning of 11,000 people in England and Wales, reported that the average household can expect to be burgled once every 40 years – although there were higher risks in various 'sub groups': for instance, in inner-city areas, houses were burgled on average once every 13 years. But an Occasional Paper detailing a survey carried out in Bedfordshire by the Christian Economic and Social Research Foundation and covering police records for 1977 and 1978 reported that, in broad terms, it must be accepted that four out of every five people living in that county all their lives could expect to be wounded, burgled or otherwise violated in person or home. And, the survey claimed, in London and other densely populated areas, the risk was much greater – persons living in them should expect to suffer twice during their lifetime.

The official Home Office criminal statistics for 1982 showed that there were 406,000 recorded burglaries, but it is generally accepted that for every one recorded there are probably two that are not. Based on official statistics, it is calculated that there is a burglary of a home in the UK every two minutes.

Each and every victim of these crimes is a casualty either physically or psychologically. Whatever the theorists and the statisticians may say, a fear remains with every victim, and with their relatives and neighbours. It is a fear that is very real, especially in some inner-city areas where people have been the victims of burglary and violent assault in their own homes more than once. In some high-risk, high crime rate areas there is little anyone can do to keep the burglars out and, even when premises have been properly secured, thieves have still forced their way in. But very often, especially in the suburbs and in urban areas, that is not the case.

The British Crime Survey found that low levels of household security prevailed generally. For example, in terms of the basic security measures of closing windows and locking doors when leaving a house empty, 22 per cent of the average household sample used for the survey admitted to leaving at least one door or window insecure on the last occasion they had left their home empty during the day. Similarly, 22 per cent of victims burgled during the daytime when their homes were empty had left a door or window insecure. But it is difficult to draw any firm conclusions from such statistics because comparison of security in victims' homes and those in the general household sample did not suggest that as a group victims' homes were less well secured than houses generally. In view of Home Office and police campaigns for improved home security measures, the findings of this Home Office research are strange, for whatever such surveys show statistically, the crime prevention campaigns are based on criminal statistics which substantiate a serious crime situation demanding a much improved level of home security.

There is evidence to show that the value of goods likely to be found in certain households might influence the burglar's choice of a target – as might be expected. In a 1976 study of the crime patterns in Sheffield it was found that the houses most vulnerable to burglary were dwellings with the highest rateable value, which might be expected to contain a large number of valuable items. In that survey, it was found that 69 per cent of victims lived in high rateable value houses.

But today, when homes in decaying inner-city areas also contain valuable items such as video recorders and hi-fi equipment, and where houses and flats in a Greater London suburb are more highly rated than homes which are bigger and more luxurious in attractive country areas, this sort of equation means little and once more demonstrates how misleading statistics can be. There are more burglaries in a month from homes on some council estates in inner-city districts than in the same period from homes in some stockbroker-belt houses in the home counties. According to the General Household Survey for 1980, published at the end of 1983, rented flats occupied by unskilled manual workers are far more likely to be burgled than the homes of middle class owner-occupiers. The survey, published by the Home Office, shows that 72 per 1,000 householders in private rented accommodation were burgled during 1980, compared with 21 per 1,000 for

owner-occupiers. The highest burglary rate reported was among unskilled manual workers. Although owner-occupiers and professional
workers were found to be more likely to report the theft to the police,
the survey found that 20 per cent of people who had been burgled had
failed to report the fact.

Of course, none of these statistics in themselves can show how this
situation is related to security measures. For instance, the group which
is normally less conscious of security and which takes the fewest
security precautions is the unskilled manual workers. So if crime
prevention campaigns can make a real impact on this large section of
society and lead to their introducing improved security, especially in
the form of better locks and bolts on windows and doors, this might
have a very big effect in reducing residential burglary.

The important thing to recognise is that the police by themselves
cannot stem the rising tide of burglary. Furthermore, they admit that
the case-load arising from the increase in crime is so great that inves-
tigative resources, especially manpower, are overstretched and the
clear-up rate is low. They, the Home Office, the Home Secretary
himself, in fact everyone connected with the maintenance of law and
order, have stated unequivocally that the public must play an important part in the battle to beat the burglar.

Whatever surveys may purport to show, there is no doubt that
increased security is an important part of that battle. Usually, it is
only after the event that home-owners are brought face to face with
reality and realise that it is their own fault that their homes have been
burgled due to the lack of even the most elementary forms of security.
It does seem that at last the sustained crime prevention campaigns
promoted by the police and the Home Office have caught the attention
of the home-owner and, as a result, the demand for security products
is growing fast. Although many more house and flat dwellers are
fitting better locks to windows and doors, only five per cent or so of
the home-owning section of the community have so far had their
premises equipped with burglar alarm systems.

Until a comparatively short time ago, neither architects, builders
nor local authorities considered the question of security when planning, designing and building homes, either in the private sector or in
the development of council estates. As a result of the rising wave of
burglary and vandalism, security demands have been forced on them
but this has meant the addition of expensive measures which could

have been included much more cheaply and effectively at the planning stage.

There have been cases where every lock provided for homes on large estates were such a cheap product, openable by the same key in every case and offering no security at all, that they have all had to be replaced. Incredibly bad design creating long, unlit walkways, dark stairs and other security hazards have made sprawling blocks of dwellings and flats a paradise for criminals. That situation is now being remedied and, with the co-operation of all concerned, from the initial planning stage to the actual construction of dwellings, security measures are now being incorporated in the construction of homes. Evidence has shown that the actual layout of new housing sites may play a major part in reducing burglary, and new home buyers will be well-advised to take a careful look at this aspect. For instance, many housing blocks were originally designed with uncontrolled access from a number of points. This meant that anyone could get in at any time for any purpose, and naturally came to be regarded as a major cause of crime and vandalism problems in tower blocks throughout the country.

Great improvements have been taking place since 1982, when the Technical Policy Division of the Greater London Council's Department of Architecture and Civic Design, after consultation with the Metropolitan Police, produced a detailed design guide for security in housing blocks. It laid down guidelines for the installation of telephone entry systems and the structural changes needed to accommodate them. One important aspect of this, recognised by the GLC, is the need to have careful consultation with the residents because the success or failure of such entry control systems depends as much on the understanding and acceptance of them by the users as on the quality of the installation itself.

The first part of a new British Standards Institute guide covering the whole field of the security of residential buildings is now in preparation, and it is hoped that it will be published in 1984. It will be the result of years of work and research by the Home Office, while the findings of a BSI sub-committee and the analysis of public comment following a draft guide issued in June 1983 have had to be assessed before final preparation. The guide will include detailed advice on all aspects of security in dwellings, including the design and layout of estates, as well as individual homes.

Open plan estates, where access at the rear is across an open area or across gardens with no high fence to provide cover for any potential intruder, are far less likely to be burgled. If there is anything a thief avoids it is operating in conditions where he may be seen and recognised or where his presence may create an alert and lead to the arrival of the police.

Many local authorities have been, and are, spending thousands of pounds on the introduction of security measures for their tenants. Electronic technology is being used by the Hammersmith and Fulham Council in London to prevent vandals and other criminals entering tower blocks in the area. Their experiment, which began on a small scale in 1978, was so successful that it has been extended to over 500 flats in three huge tower blocks in the adjoining area of Shepherd's Bush. The scheme involves the installation in the flats of entryphones which are linked to a closed circuit television system, enabling the resident to see on his television screen exactly who comes in and out of the electronically operated main doors. Fears that council tenants might feel they were being spied on were soon dispelled when a survey showed that 97 per cent of them approved of having a locked street door and 92 per cent approved the use of the television equipment.

Wandsworth Council has been installing an even more sophisticated entryphone system linked to a closed circuit television system in a number of tower blocks. All tenants have been given special security keys, and the closed circuit television system provides a special screen display which shows both front and back doors at the same time.

These entryphone systems have been successfully introduced into many council homes in other parts of the country, such as Manchester, where very high crime rates have been recorded. And Merseyside housing associations joined forces with the police three years ago in a campaign to reduce the number of break-ins and to give better security to their tenants. They produced a booklet outlining their basic security requirements to be applied to all new housing association schemes.

Recognising the importance of incorporating security measures into homes at the planning stage, Avon and Somerset Police appointed the first-ever architectural liaison officer in their crime prevention department. Within a short time this move achieved an important breakthrough in the provision of security for the private home. Ladbroke Homes agreed that all new houses built by them would be fitted with good quality locks conforming to British Standard 3621; that all front doors would be equipped with a safety chain; window locks would be fitted to all ground floor windows; and each house would be equipped with a 'bells only' alarm system. The company also decided to offer house buyers the option of having a safe specially recessed into a floor cavity, which would be free of charge except for the cost of the hardware.

A second company, Wimpey Homes, agreed to offer a general security package to home buyers purchasing any of their homes in the West of England, which included a British Standard mortice deadlock on the front door in addition to the existing nightlatch, a front door safety chain, and locks on downstairs windows.

The appointment of the specialist architectural liaison officer in the crime prevention department benefited Bristol City council house tenants as well. The council had decided to replace old metal windows and wooden doors with aluminium ones because of their long-term effectiveness, and recommendations on their design and style made by the liaison officer ensured they incorporated security against burglars. The simplest method would have been to fit standard window locks. But people often forget to use the locks, or leave a window open for ventilation, so the architectural liaison officer decided it would be wiser and more effective to design windows that would prevent unauthorised entry under all circumstances. Building regulations stipulating minimum areas for opening vents had to be complied with – a fact which is often unknown to, or disregarded by, the private home-owner carrying out his own D-I-Y window replacements – and it had to be possible for all sliding windows to be cleaned from the inside. The recommendations made to Bristol Council were accepted and resulted in a design which ensures that the windows cannot be opened more than six inches unless there is 'positive' action, such as operating a lock, by the occupier. It is possible to leave the window open, therefore, for ventilation and yet still to prevent entry.

The principles upon which the recommendations were made should be borne in mind by all home-owners looking for improved security:

1. Burglars hate noise and will rarely break a window if they can help it.
2. The burglar will use methods such as drilling or cutting a small

hole in order to manipulate the catch inside and then open the window if it is not locked.

3. If the burglar does break the glass to reach the catch, he will rarely risk climbing through the jagged edges if he still cannot open it.

4. The larger the pane of glass, the less likely it is to be broken.

A number of other police forces, including those in Nottinghamshire and South Wales, have now appointed architectural liaison officers and it is expected that all other forces will follow suit in the near future.

Some well-known house building organisations have themselves recognised the growing need for in-built security. For years now the building industry has been criticised for generally fitting grossly inadequate locks in new homes, despite the fact that the extra cost involved would be minimal. And until now they have completely shied away from the suggestion of incorporating any provision for security systems, on the grounds of cost. Home buyers are themselves very much to blame for, if they were to demand that a certain level of security should be part of the design and construction, the position would quickly be remedied. But we are back to the same old story – few house buyers think of security and do not recognise the risks which exist.

Among the companies which have decided to offer an option of in-built security to new home buyers are Wates and Barratts. In the case of Barratts, the option is offered only on a range of very costly premises far beyond the reach of most people. But Wates are offering the option of a complete home protection package by Chubb at a mixed development site in at least one area, and exhibition houses are equipped with the system for the inspection of customers. One company has tried this idea of offering a particular package before but it has not been successful. With an arrangement with one specific company, it does mean that the customer has no choice. Most home buyers prefer to buy premises with a basic in-built security in the form of good, strong doors, window locks, and perhaps the wiring necessary to install an alarm system – leaving the decision and choice of a system open to the individual concerned.

What every home-owner should remember is that there is a free independent advisory service available whose advice is completely dependable and impartial: the local crime prevention officer. All you have to do is to contact him through your local police station and he will call to survey the premises and make whatever suggestions he thinks are necessary to improve security. This advice is vital, for he is an expert, with a working knowledge of every kind of break-in method, and he can recognise the risks and suggest the right way to deal with them.

It is now becoming increasingly obvious that the blanket approach to crime prevention and security is not the best – there are local situations which differ from others and call for different precautions. It must always be remembered that, contrary to popular belief, the greatest number of burglars are not professional operators but opportunist thieves, and even simple precautions such as fitting good quality locks will deter them. Depending upon the environmental and geographical circumstances, the nature of the premises and taking into consideration the financial position of the home-owner, it may not be necessary, therefore, to install an elaborate alarm system – or, indeed, any alarm at all. However, such a system, whether it be elaborate or simple, does help to overcome the fear of crime by ensuring peace of mind and giving reassurance.

People living in certain areas, such as inner cities, face a much greater risk since their homes are more vulnerable, especially when unoccupied, and therefore need more protection. The doors of many flats, especially rooms in converted houses, are often of very poor quality and entry from deserted corridors by simply forcing the door or picking the lock is simple, especially during the daytime when the flats may be unoccupied. Special protection is required in multi-occupation blocks and converted premises where open doors to the street allow anyone to gain access and where there is no possibility of surveillance. Burglars do detest noise and loud alarms may drive them away. Strong, reinforced doors with really good locks and hinge bolts give a fairly good level of security.

The threat of burglary is just as great to homes situated in remote districts. Such homes may be sited some distance from the nearest neighbour, or stand in their own grounds hidden from the view of both neighbours and passers-by. These are the type of premises most in need of a sophisticated alarm system which will supply round-the-clock protection linked to a central station so that any alarm is at once relayed to the operator.

The introduction by police forces all over the UK of 'Neighbourhood Watch' schemes is probably one of the most successful methods of making a whole community security-conscious, while at the same time obtaining information which will help the police both to identify and capture criminals and frighten potential burglars away. Based on a highly successful American experiment, the idea was first introduced in a stockbroker-belt village in Cheshire where there had been a spate of burglaries. The scheme has now been extended into urban and rural areas of all kinds, into densely-populated London districts, and is operating in areas containing every type of housing.

Neighbourhood Watch provides the framework within which a community can be to some extent self-policing in a strictly preventive sense. It has been made clear that neither vigilante groups taking the law into their own hands nor prying busybodies are wanted, and applicants are carefully vetted before the schemes are set up. It means that members of a community are organised to work together and with the police to keep a constant watch for anything suspicious, to report to the police when necessary and, by identifying potential risk areas in their neighbourhood, to devise any necessary measures to reduce criminal opportunity.

An integral part of the scheme is that trained officers carry out surveys of homes and point out security weaknesses which should be rectified. In high-risk crime areas and in densely populated districts of inner cities, the constant observation by neighbours of other people's property and homes will, it is hoped, bring about a big reduction in crime and lead to the quick arrest of the villains when a crime is actually committed.

However, perhaps the most important aspect of the introduction of a Neighbourhood Watch scheme is that, in the areas where it is in operation, window stickers and street signs are put up announcing the fact. Few criminals will make a move if they think there is the slightest risk that they will be spotted, and as soon as they see the Neighbourhood Watch signs they know that the ordinary people who will become their victims are watching and ready to report anything suspicious. Experience has already shown that this alone causes most criminals to keep away. It is true that this may mean merely that criminals are being deterred from working in that area and will at once transfer their activities to another. The answer to this must be for the growth of the Neighbourhood Watch scheme to be encouraged until every

area and every block of flats is covered. But until there has been
sufficient time to assess the results achieved so far, there is little or no
evidence to indicate that any crime reduction in the areas with home
watch groups is offset by a proportionate increase in areas without
them.

One thing is quite certain – the evidence so far shows that in most
of the areas where such a scheme is in operation there has been a
significant reduction in crime, especially burglary. When an experi-
mental scheme was introduced in a crime blackspot on a Bridgend
housing estate, burglaries stopped; when the project ended, they
started again. Unfortunately, success often brings complacency and
then apathy. In the United States, where the scheme was first developed and where there are many of them in operation, initial results
were amazing. In Detroit, a crime blackspot, the areas which introduced Neighbourhood Watch schemes experienced not only a
dramatic reduction in the overall crime rate and the number of home
burglaries falling by 61 per cent, but also a big reduction in the fear of
crime. In every case, success could be attributed to the commitment
of the householders and other members of the group and the crime
prevention officers. After about three years of low crime figures,
however, a sense of false security caused residents to think they no
longer needed to take care and the crime totals began to rise again.

There is no doubt about the value of this kind of policing based on
the partnership of police and people, but it must not allow home-
owners to ignore the fact that, in addition, there must be adequate
physical protection of the home and the family. That is why over the
past two years the Home Office has spent more than one and a half
million pounds on a nationwide campaign to encourage everyone to
fit window locks. The campaign has met with great success and the
sale of both window and door locks has rocketed. It cannot be stressed
too strongly that it is this basic protection which must be the first
priority. With it goes the importance, which once again so many
people overlook, of door safety chains so that an intruder on the
doorstep cannot easily force his way in. Most of the sophisticated
alarm systems may be beyond the pockets of many people, especially
pensioners, but window and door locks and door safety chains are
simple and effective deterrents and are also cheap. Those wondering
what to give their elderly relatives as a birthday or Christmas present
might well consider buying these deterrents for them.

2 Locking the burglars out

Although two-thirds of all burglaries take place by entry through windows, only one in ten homes have window locks fitted. Of course, nothing will stop a really determined thief provided he has the time and the opportunity to overcome obstacles, but locks, bolts and bars will deter him. It is important to remember that the longer it takes to break into premises, the greater the chance of detection before any real damage is done. Security measures buy time if nothing else, and time is the enemy of the would-be intruder, who always looks for the easy way in – and out again. If it is not easy many thieves, especially the huge army of opportunists, do not waste time or chance their arm on making a lot of noise and attracting attention. They will go away and look for a target where there are no safeguards. So why make it easy?

A recent campaign by the Home Office to encourage householders to fit window locks highlighted the fact that most burglars come in through the window – usually rear windows which are hidden from the view of neighbours and passers-by – and resulted in a 43 per cent leap in lock sales in the selected areas in which the campaign was concentrated. This was followed by another similar campaign on a national scale. Nevertheless, crime prevention officers are appalled by the large number of people who still do not take any precautions to secure their windows, although the necessary measures are both simple and inexpensive.

A wide variety of locks and catches is available for security fitment to timber-framed windows, whether of the casement, fanlight or sliding sash type. In addition, locks are available to secure most types of steel-framed windows, and the latest developments are designed for the aluminium windows now being fitted so extensively as replacements. Since aluminium is a soft metal which can easily be forced, the only way to ensure security is to fit one of the locks produced solely for this kind of window. Many manufacturers are now including

Window lock: the circular nut is screwed down by hand and when it reaches the end of its run, it 'free-wheels' and cannot be removed except by using the special key provided.

Left
Lockable casement fastener: the disc-tumbler lock prevents the fastener being operated.

key-operated locks at the production stage and buyers should make certain that such security is incorporated. Key operation is absolutely essential, for once the householder has locked the device and removed the key it ensures that an intruder breaking, cutting or drilling the glass to reach the catch inside will still be unable to open the window.

One of the greatest gifts to the burglar in recent years was the introduction of glass louvre windows, especially when they are used, as in some houses, to glaze an easily accessible and large aperture. It takes only a matter of seconds for a burglar to remove the panes of glass from their flimsy aluminium frames. Home Office advice to occupiers with this type of window has for some time now been to use a good epoxy-resin glue to stick the slats of glass into the frames: *not* the so-called superglue which will not prevent removal of the glass by sideways force. Home buyers should bear in mind the risk attached to these glass louvres and remember that neither the Home Office nor the police recommend their installation. However, there is now good news for those who have them. Design engineer Paul Lehmans, who carried out his own research, has invented a simple and revolutionary lock which makes louvre windows as secure as any fixed window. In a simple two-stage operation the glass slats are cemented into the casing and the new lock installed to secure the catch handle on the inside. The lock has been designed so that it can be adapted for any type of louvre catch, of which there are many different kinds. Crime prevention officers from all over the country have been to the London premises of the Louvre Lock Company to see the new device.

The bewildering array of locks and bolts available to the householder these days causes great problems of choice. Many of them are general function devices and many are designed for special tasks. There are still far too many people who think that all they have to do is to buy a lock, fit it, and that everything will then be safe and sound. But nothing could be further from the truth. Very careful thought is needed in the choice of a lock to make certain that it serves the purpose for which it is intended. There are cheap and flimsy locks which are a waste of money and are useless from the security point of view. Good lockmakers make good locks although, unless there is some very special reason, this does not mean they need be very expensive.

Research is always going on and the lockmakers have come up with some significant advances in design, but the basic principles are still the same and all else is a variation on a theme. It cannot be impressed upon householders and flat dwellers strongly enough that when they think of door locks, they must think of the doors first of all. There is one principle to keep in mind: namely, a lock is only as strong as the door to which it is fitted, and the frame into which the door itself is fitted. That is because all conventional-type locks have the same weakness – they shoot a bolt in one direction only and at one point only.

Most doors and frames are still made of wood, often of poor quality, and the thief who is faced with a good quality lock in a poor quality door frequently disregards it completely, especially if the premises are in a situation where making a little noise will not attract attention. He attacks the hinge side instead and either lifts the door off the hinges – and remember, outward-opening doors are particularly vulnerable in this respect – or, if the door is flimsy or the hinges are poorly secured or worn, uses a jemmy to force it away from the frame. Fitting hinge bolts will give protection against this. Even if the door is a really good one but has exposed hinges, it should always be fitted with hinge bolts to prevent burglars unscrewing them. Hinge bolts are simply fitted by drilling suitable holes and placing them a few inches from the hinges. One bolt is not sufficient – they must be in pairs: one near the top hinge and the other near the bottom hinge. The bolt is driven home into the hole drilled in the closing edge of the door. The locking plate is recessed into the frame, so when the door is shut it gives great strength to the whole door as well as protecting the hinges. Here, again, there are different types available, so make certain they are manufactured by one of the reputable firms.

All house and flat dwellers should make certain that exit and entry doors are always of solid construction and never of the 'hollow' or 'egg-box' type in which thin coverings such as plywood or even hardboard are fixed over wooden battens so that the interior is a series

Hinge bolt designed to secure the hinge side of the door, especially if it is of a type where the hinges are exposed. Hinge bolts should always be fitted in pairs.

of empty chambers. It is obvious that these have no ability to withstand attack. Thieves merely cut holes, smash in one or more of the panels, or simply kick them in. Any entry and exit doors which have deteriorated through age or wear should be replaced, especially if there are wide gaps between door and frame into which a burglar can easily insert a bar to force the door open. No amount of money spent on a lock will make a weak or rotten door safe. Few home-owners realise when replacing entry and exit doors that, if they are prepared to spend a little extra, they can choose a type strengthened for security. Special burglar-resistant door sets are available from several manufacturers and distributors. These do not look unsightly and cannot be distinguished from wood, but the doors have a steel plate in the centre and reinforced frames.

Finally, attention must be paid to door frames. However good the door and the locks may be, if the frame is weak and fixed to a thin plaster-faced wall, especially if wear means that the wall itself is broken and cracked, it negates the value of the door security and intruders will often wrench the whole door and frame out. Frames can be strengthened by the use of metal bars and it is sometimes advisable to have this done. The local crime prevention officer will always give advice.

Home Office and police advice is that a safety chain should always be fitted to the front door and should *always* be used when the door is opened to a caller. But this does not mean just any chain: cheap ones offer no real security at all as the fittings securing them can easily be wrenched out by a powerful thrust against the partly opened door from the outside. The safest types are those which have really strong chains that are not too long – otherwise an intruder might get an arm through to grab the occupant – have very stout fittings both to door and frame, and have strong fixing screws.

There are stronger and, in some respects, much better front door safety devices on the market today which are well worth considering and may be a better investment for some types of dwellings. They are 'limiters' which cannot be released by someone jerking the door open and shut because they operate by means of a solid bar. Ingersoll have launched one of these, called the Door Check DSC2. It consists of a sliding arm attached to the door frame which engages in a strong metal bracket fitted on the door itself. When the door is closed, the bolt action of the sliding arm provides a positive action which adds to the general security of the door. Sideways movement withdraws the bolt and allows the bracket to make a restricted movement along the sliding arm. Once a caller has been identified, the door is temporarily closed and a simple pivot action disengages the sliding arm to permit the door to be fully opened.

For the protection of the final exit door when the premises are occupied – these can be bolted only from the inside – and for added protection on all other doors of exit and entry as well as wooden windows, key-operated security bolts which are morticed into the door or window frame are used. They are invisible from the outside and the bolt goes into a striking plate over a receiving hole in the frame. Two are fitted to each door or window: one at the top and the other at the bottom.

The popularity of the aluminium patio door has presented parti-cular security problems, because this material is soft and malleable. It has made entry by an intruder much easier, quicker and quieter. Con-ventional locks could not be fitted in the older types – which are still installed in many thousands of homes – because of the method of construction and the relative thinness of the door stiles and frames. The risks which these doors posed was first realised in the United

Door (*top*) and window security bolts which are morticed into the door or window frame to provide additional security.

States more than 30 years ago, and the internationally known lock manufacturing company of Adams-Rite researched the problem in California. They came up with the idea of the swing-bolt lock, which hooks over a keep recessed into the frame, thus preventing it from being forced back. This type of lock has since been fitted to aluminium sliding patio doors all over the world and is still in common use, although modern versions have much improved on the original design. Hundreds of homes still have patio doors which are vulnerable to attack, and entry through such doors is commonplace. But protection is now both simple and inexpensive and every householder should make sure that the vital security measures are taken. Blocks which are screwed into the top channel are designed so that they do not impede free movement but prevent the doors from being lifted out of the bottom channel. And there is now a big variety of special locks available which have been designed specifically for patio door security. Some are centre locks which are key-operated and are fitted by means of drilled holes through the centre of the frames of both doors where they meet in the middle in the closed position. Some types are screwed to the edge of the door itself on the inside and move with it when in the unlocked position. One new type, developed by the West Midlands firm of Fullex Windowcraft, features two adjustable steel pins with locking nuts as the locking points and a pin tumbler deadlock which can be fitted on either the inside or the outside of the door. It has been designed to be fitted at the production stage.

The ideal way to obtain maximum security at minimum extra cost is, of course, for the locking system to be included in the manufacturing process, and double-glazing companies are beginning to introduce units with such safeguards. Some are going even further and incorporating either a burglar alarm system or sensors which may be used to link into any burglar alarm system which may later be introduced into the home, or which may already exist. All home-owners are advised to take a careful look at as wide a range of units as possible before deciding on any installation, and to demand that there are built-in safeguards. One company which has included both aspects in its latest range of high-security patio doors is Monarch Aluminium Ltd. A special digital intruder alarm system is incorporated during construction and it is set by means of a code devised by the house-holder – a clever security move for the protection of the code itself – who keys into the system using a series of numbered buttons. A major

Patio door lock.

feature of the doors is a multi-point locking system constructed in stainless steel, which ensures that the doors cannot be levered open from any side. Monarch, incidentally, also supply high-security residential doors which incorporate the digital alarm system, and have special security glazing and either a cylinder mortice lock or a night rim latch which can be deadlocked from outside or inside.

The danger of fixed aluminium double-glazed window installations and fixed secondary window units, following a number of deaths of adults and children trapped inside their burning homes, has highlighted the dilemma which has always faced the security expert and the fireman. Effective security entails keeping the intruder out, while the fire hazard requires the provision of a fast and easy means of escape for the family if necessary. Attempts to satisfy both these requirements have not always been successful, and the escalation of burglary has inevitably meant that much greater emphasis has been placed on the provision of anti-intrusion measures. Unless the situation is carefully viewed as a whole, however, either or both aspects may fail. Inevitably, unless advice is taken, the greater security a house or flat occupant achieves, the greater is the risk of being trapped inside if fire breaks out.

In the event of a fire, even if there is a possibility of reaching patio doors protected by the type of added-on security locks described earlier, there is going to be delay while the locks are opened, especially those of the centre-fitting type. Thick smoke and fumes may even hamper finding the key – and it takes only a very short time for someone to be overcome and rendered unconscious. This, in my view, is a criticism which can be levelled against security bolts which are recessed into windows and doors, especially if, for maximum security, they are installed both at the top and the bottom, and are operated from the inside by a key. Not only are those on the front door, which is usually the final exit door, of value only when the premises are occupied – the least risk period – because they can be bolted only from the inside, but there is the question of considerable delay in the event of any emergency such as fire, because it is necessary to carry out three unlockings – the main lock and the security bolts, needing two different keys – before an escape can be made.

As far as patio doors – and other double-glazed units – are concerned, some manufacturers are now using uPVC material instead of aluminium but this, again, has posed security problems. This material may be jemmied or sawn through unless steps are taken at the manufacturing stage to include reinforcement. A range of double-glazed units designed and manufactured by ComforTec Windows Ltd are made of uPVC and the company has taken unusual steps to try to combine maximum security with the means of fast and easy exit. The patio doors have frames reinforced throughout with galvanised steel. Concealed within the frames and distributed round the perimeter of the sliding pane are eight bolts, especially shaped to prevent any attempt at lifting the door out of its channel, which form part of a multi-locking system on all sides operated by one handle. A key-locked ventilating position allows slight opening but prevents anyone forcing an entrance. Internal glazing beads prevent the glass being removed by the use of a jemmy from the outside – another favourite method of entry by burglars who have the time to work undisturbed. Laminated glass of a standard which will withstand hammer blows is used. A foot-operated lever needs only a touch to unlock all the bolts in an emergency and the sliding door instantly glides fully open for immediate escape.

Multi-point locking systems were pioneered in the UK by the lockmaking firm of Ramicube Ltd, who produce the 'Mul-T-Lock' system now fitted to the front doors of many homes. Their locking systems fit existing doors, but the company also supplies a whole range of purpose-built doors incorporating this system. There are currently a number of such systems available. They consist of a main lock that includes an additional mechanism which, when operated by turning the key, extends bolts from either three or four sides of the door and deadlocks them into keeps surrounding the frame. Where the three-bolt system is used, the rear edge of the door is secured by hinge bolts. Jemmying is prevented because there is virtually no movement between the door and the frame. The force of any attack on the door is dissipated through the rods connecting the bolts and thus distributed via the keeps to the surrounding frame and wall. It is claimed that this system is the only real advance in design over the conventional type of lock and that it renders forced entry almost impossible, but such a system is much more expensive.

Many people believe that they add to security by locking their internal doors as well. Not only is this a grave mistake but it could have disastrous results for the occupier. Once a burglar is

Mortice five-lever deadlock suitable for fitting to the main entrance door.

inside unoccupied premises he has all the time in the world to make short work of such obstacles – and he will. Locked internal doors have been ripped and jemmied from their frames, surrounding woodwork has been splintered and, in some cases, intruders have even hacked or kicked holes through the adjoining walls if they are of the lath and plaster type. Never, *never* lock internal doors.

The most effective security for the main entrance door, which in most cases is the front door, is a high-quality mortice deadlock, which is normally fitted in addition to the ordinary existing latch lock. The term 'deadlock' means that it is of the type which, once locked, can be opened only using the correct key.

These locks are morticed, or in other words recessed into an aperture cut into the door, and the bolts are shot into a hole cut in the frame and protected by a surface metal plate. This means they are very much stronger than surface-mounted locks, which may be wrenched or forced off. There are many types to choose from, some of them having special refinements to give even greater security, but the minimum standard is that of the five-lever lock. This gives a huge number of 'differs' or combinations to prevent its easy unlocking by an intruder merely trying a small number of keys. The lock fitted by a builder is usually only a two- or three-lever model with very limited security, if any. Five-lever mortice locks have a mechanism involving anything between 1,000 and 1,500 differs. For those who want even greater security there are seven- and ten-lever types offering thousands of different combinations.

A number of different types of cylinder mortice deadlocks are also available, some of which are particularly suitable for doors made of aluminium or with very narrow stiles. Some of these locks are operated from the inside by a knob or handle and should never, therefore, be used for doors with glass panels, since a thief has only to remove the glass to open the door. This applies also to locks which are accessible through the letterbox. In such situations only locks without any internal operation or with key operation should be used. But cylinder locks like this have some advantages over lever locks: the key is much smaller and the permutation on differs ranges from 30,000 up to millions.

Ordinary rim locks, which many people still have as their only insurance against intrusion, have no security value at all. The high-security rim lock is of the automatic deadlock variety and anyone with

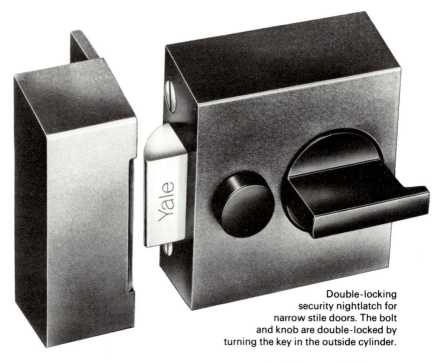

Double-locking
security nightlatch for
narrow stile doors. The bolt
and knob are double-locked by
turning the key in the outside cylinder.

the ordinary rim nightlatch lock can replace it with one of this type. They are, of course, not as secure as a mortice lock because they are surface-mounted but, on the other hand, wooden doors, particularly those with narrow stiles, are to some degree weakened by the cutting away of wood for a mortice. A great deal of research has gone into the design of modern rim deadlocks and the safeguards which are built into them to prevent interference. For instance, Chubb have recently launched a new automatic rim deadlock for front doors which has a 50,000 variation, an unusual key shape to guard against duplication, hardened steel plates to protect the mechanism from drill attack, in-built protection against wrench and torque attack, and a lockable interior knob designed for easy use by arthritic fingers.

Some lock manufacturers offer a key registration system which can be useful, since it is not possible for anyone to merely walk into a key cutting establishment and have one made. Only the registered owner can obtain additional keys and he has to order them from the manufacturer. Very often the lock itself is of a unique design to prevent duplication.

There are very high quality security locks – often recommended by insurance companies for premises where there are above-average risks and value of contents – which can form part of a master key scheme linked to alarm systems fitted with micro-switches, but these sophisticated methods are expensive and probably only of real value, especially in cost-effective terms, to very large residential premises or in exceptional domestic situations. However, one company, Kaba Locks Ltd, has recently designed a heavy-duty mortice deadlock which is suitable for the final exit door in the home and offers the option of a micro-switch which is ready wired and, actuated by the bolt, can operate lights and control signals or be linked to an intruder alarm. This sounds a very attractive new development.

Both Kaba Locks and Salsbury Locks Ltd offer a master key system for the home, which gives the convenience of a one-key house if desired. Both systems offer the ordinary home-owner, especially those living in detached houses in urban areas, a lock security facility so key holders can have a complete range of the locks fitted which can be

Front door lock featuring a lockable interior knob to prevent an intruder opening it even after breaking the glass.

operated either by different keys, or by one key only. This choice is an attractive one, since opting for one key means that there is no need to carry around bulky and heavy bunches of keys on a ring.

After four years of research, Salsbury Locks have invented a twin-bladed key and a range of 'instant differ' locks, and their locking system gives a facility which not only means one of their unique keys will open all the doors in both home and office, but the padlock on a bicycle as well! Here again, the home-owner is provided with an attractive new development. The twin-bladed key has four different edges cut in two planes and can be cut to four million different combinations to fit a compatible instant-differ cylinder lock. The key holder's unique combination and his signature are registered on microfilm records so that no one else can get a copy cut without his agreement.

Lock design is being developed all the time. All the big manufacturers such as Yale, Copydex, Legge, Chubb, Ingersoll, Banham and

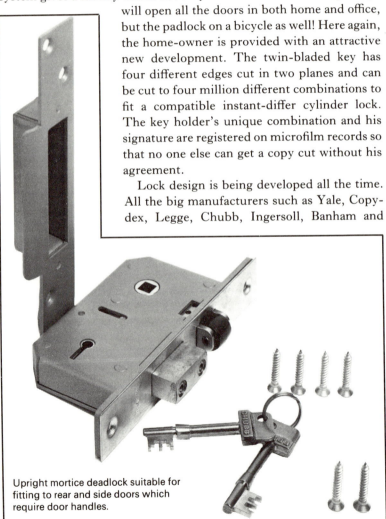

Upright mortice deadlock suitable for fitting to rear and side doors which require door handles.

Castell, to name but a few, produce locks with unique features claimed to protect against all forms of attack and duplication. Abloy Locking Devices, who also offer a master key system, have developed a unique lock cylinder using rotating brass discs which has been incorporated into a range of more than 150 different basic lock types.

Many people forget that most burglars obtain entry through rear doors and windows. Back and side doors, which are not normally used as final exit doors – if they are, then their lock security should be similar to that of the front door – should be secured by top and bottom surface-mounted bolts of a very robust type, or by key-operated security bolts, in addition to a rim or mortice deadlock of British Standard 3621. For the benefit of the uninitiated, this means that the lock will bear the familiar 'kitemark' to indicate that the lock has been manufactured to the standards laid down by the British Standards Institute. This states that the lock should feature at least five levers, 1,000 key variations and a mechanism protected by special plates which resist drilling and physical attack.

All double doors should have bolts at the top and bottom of both doors, as well as a lock. Always remember to shut and lock all windows and doors whenever you leave home, even if only for a few minutes: that is all the time an opportunist thief needs to nip inside and quickly grab a handbag, loose cash or radio before escaping. And it is during *daylight* hours that most housebreakings take place.

No lock is worth anything if keys are left lying around. Never leave keys in the lock – always take them with you.

Never leave keys in hiding places such as under the mat or inside the letterbox – thieves know just where to look. Allow members of the family to have spare keys – or leave one with a trusted neighbour. And whenever you move home remember that a lot of people may have keys to fit the locks of your new residence, so spend a little extra and have new locks fitted throughout your home.

Two final but important points: pay special attention to the security of windows which may be reached by climbing a drainpipe or clambering on to a flat roof – these are entry routes which burglars look for. And remember that burglars know that in the evenings most families assemble in one room to watch the television and that the sound will drown any noise they make. So shut and lock all doors and windows before you settle down to an evening's viewing.

3 Alarm systems

The design and manufacture of intruder alarm systems for the protection of the home is probably the most active sector of the security industry. More and more established manufacturers who previously produced alarms exclusively for commercial and industrial premises are now adding units specifically created for the domestic market to their range, while new companies are springing up to serve just that market. This development is bringing with it a wider and wider range of devices and systems to confuse the prospective buyer. Technical developments, including the use of the micro-chip, are leading to growing sophistication and the elimination of problems which, until comparatively recently, had evoked adverse criticism of the domestic alarm field.

A few years ago, the cost of installing a burglar alarm system deterred most people, but prices have been dramatically reduced now to a level which suits all pockets. Systems range from simple and maintenance-free units which cost as little as £50, through an extensive middle bracket priced between £400 to £600, to very sophisticated systems in the £600 to £1,000 range. One area of the domestic market which has proved successful is the 'do-it-yourself' field, the chief attraction of which is the relatively low cost since there are neither installation nor maintenance charges to pay. The lower end of the market has been invaded on a massive scale by nationally known household equipment manufacturers such as Hoover and Pifco. Hoover, for example, under the name of 'Thiefcheck', have introduced a complete home protection package which is available from almost all D-I-Y, hardware, ironmongery and electrical shops and department stores.

Great care must be exercised by potential customers in the choice of a system. The market is very competitive and extravagant claims are sometimes made by the less reputable manufacturers and suppliers. Some domestic products which have been heavily advertised have failed to make any impact and have disappeared from the shelves. In some cases customers with professionally installed systems have been disappointed both by results and maintenance failures.

One of the weaknesses of the single-type do-it-yourself security packages is the fact that they may be installed in situations for which they are unsuitable. This has already been recognised by one of the manufacturers of D-I-Y security systems produced for the mass market. Philips Service offer three home security systems, each of which is designed for specific applications. One is designed for three- and four-bedroomed houses and includes a personal attack button and a self-activating alarm unit with a battery, which will sound even if the cable to the control box is destroyed. The second, which does not incorporate so many features, is designed for the two- to three-bedroomed house. The third system, on a still smaller scale, is for flats.

Professionally installed systems, although much more expensive and usually carrying a maintenance agreement which adds to the cost, are usually tailored to meet the particular requirements of both the customer and the type of premises. Few people realise that a system which may be very satisfactory in one set of circumstances may be useless, or only partially effective, in others. The structure and layout of residential premises varies greatly, as does the geography and environment. Even the degree and the nature of the security demanded varies from customer to customer. All reputable companies supplying units which need professional installation employ trained surveyors who will visit the home and give expert advice as well as adapting the system to the premises. But remember, they are working for the company from which you are buying the product and this survey cannot give you any indication whether another product or another type of system might be better. The biggest problem facing the home-owner is obtaining independent advice before making any decision. However, it is there for the asking from the local crime prevention officer and, although he cannot recommend any particular product, he can explain the different systems and suggest the type which is the most suitable. Alternatively, pre-installation advice may be obtained from the National Supervisory Council for Intruder Alarms.

Unfortunately, there are few opportunities to examine and compare the whole range of systems and makes available. There is one quarterly magazine called *Home Security and Personal Protection* which is on sale to the public, and is the only one of its kind in the UK dealing

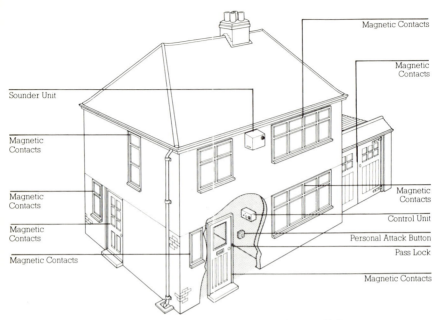

Magnetic Contacts

Magnetic Contacts

Sounder Unit

Magnetic Contacts

Magnetic Contacts

Magnetic Contacts

Magnetic Contacts

Magnetic Contacts

Magnetic Contacts

Control Unit

Personal Attack Button

Pass Lock

Magnetic Contacts

Layout showing sitings of components of a typical do-it-yourself alarm system.

exclusively with the subject. There are some fairly comprehensive exhibitions from time to time in centres such as Birmingham but the biggest is an annual event at Olympia. Called the International Fire, Security and Safety Exhibition, it is basically a trade show but attracts thousands of home-owners because it provides an opportunity to inspect the entire range of domestic security products, both British-made and foreign, which are available in the UK.

Although there are many different systems, they all have one common aim: to detect an intruder, react by raising an alarm and then deter him by the use of light or noise, or both. There are several different methods of detection. Some of them require simple wiring circuits to connect sensors – for example, from window to window. The wiring may be buried in the wall or run round skirtings and window frames. The sensors, or detectors, may be magnetic contacts, pressure pads or micro-switches. The magnetic contact, which is a switch, is most commonly used. The contacts are kept closed by a magnet attached to the moving part of a door or window so that, when the magnet is moved away by opening the door or window, it causes an alarm to sound. But if the window is broken instead of being opened, this does not break the contact. To protect against both methods of intrusion a vibration detector is used. This is a switch kept in place by a weight which is momentarily dislodged by the vibrations of an attack, thus triggering the alarm. A weakness of magnetic sensors is that burglars have found that, by using a stronger portable magnet, they can keep the contact closed while opening the window and can therefore prevent the alarm sounding.

Pressure mats are sensors which are placed beneath carpets and, when they are trodden on, two pieces of foil are pressed together to make a contact which will then sound the alarm. They are usually

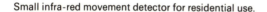
Small infra-red movement detector for residential use.

sited in places where an intruder is likely to step, such as under windows or at the bottom of the stairs. But they pose problems: they often 'creep' under the carpet and thus change their position, or a householder may unwittingly place a heavy piece of furniture on top of them which will gradually sink down until it closes the contact and therefore create a false alarm. Manufacturers and designers have introduced all sorts of refinements to remedy these weaknesses. Many window and door sensors, for example, now include anti-tamper facilities so that an alarm is activated by any attempt to negate them.

Many systems use what are loosely called space or movement detectors which are designed to protect whole areas such as rooms and corridors. Unfortunately, they have always been prone to false alarms, although research and development have ironed out many of their faults. These detectors are either passive infra-red (known as PIR), ultrasonic or micro-wave.

PIR units transmit a constant and passive infra-red pattern created by the room and its contents. The body heat of an intruder changes this pattern and triggers an alarm. This is probably the most commonly used movement detector but it has always had a high false alarm rate because other changes in the protected area can change the infra-red pattern and cause an alarm; and there are blind spots in and around the coverage area. It must be said, however, that recent developments have reduced, if not eliminated, these difficulties.

Left: Ultrasonic movement detector which uses ultrasonic sound to detect the movement of an intruder in a specific area. *Right*: An indicator unit for use with an ultrasonic detector to indicate in which area the alarm has been raised.

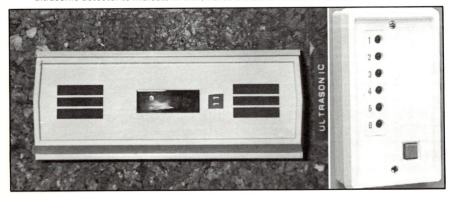

Micro-wave movement detector.

Ultrasonic devices radiate energy at a frequency which is changed when a movement takes place within the energy field. They are very reliable, but because they operate on a principle which enables them to hear sounds above the range normally audible to the human ear they, too, have been subject to a high false alarm rate. They have a tendency to be affected by forms of turbulence not due to an intrusion, such as telephone bells. Designers are now coming up with highly complex signalling circuitry to combat this false alarm syndrome by analysing the sounds and being programmed to reject those which are not made by genuine intrusion attempts.

Micro-wave detectors operate by means of a radio frequency and comprise a module which both transmits and receives a radio wave in the micro-wave band. If certain conditions are created, a relay contact opens and an alarm is triggered once the unit has analysed the signal, its strength, frequency and duration and then decided by its programming whether it has been caused by human intrusion. However, a lot of things from running water to a vibrating surface may produce the same set of conditions and initiate an alarm. And, as micro-wave energy penetrates most building materials, care must be exercised to prevent movement from outside activating the alarm.

This question of the false alarm has always been one of the big problems with alarm systems, but designers have overcome many of the causes and are always trying to make alarms more reliable. In spite of the much-publicised estimates that these false alerts account for 98

Components of a fire/intruder alarm kit.

per cent of all calls from alarm systems, there is little real evidence to show the true position. Many of them could perhaps be the result of an intrusion attempt where the would-be burglar has been scared off. And, after all, crime prevention is the important aspect.

The continual ringing of alarm bells has always been an environmental nuisance. Not only does it cause great annoyance to the neighbours, but it discredits the system concerned, and the security industry in general, by convincing many people that a burglar alarm is a waste of time. Worse still is the cost to the public purse in the waste of police time and manpower and other resources, and the tendency of many people to simply ignore all the alarms because they are always sounding off for no reason. One thing that has done much to encourage the manufacturers to introduce reliable safeguards against this failing is the decision by most police forces to withdraw response to troublesome systems after a specified number of false calls, and to restore it only after certain conditions have been satisfied.

The majority of alarm systems now use piercing siren sounders in place of the traditional bell, and most of them emit 120 decibels. This fact alone demands more and more research into prevention of false alarms. To comply with new noise regulations and to avoid annoyance to the neighbours, all modern alarm systems are designed with an automatic cut-out which switches off the alarm if the occupier or other key holder does not do so manually within a pre-set period, which may vary from three up to a maximum of 20 minutes. This latter period is important because it is the maximum time under the noise regulations that any alarm should continue to sound before being silenced. Of course, burglars know all about the cut-outs. So, in order to eliminate the hazard posed by the intruder just waiting in hiding nearby and then returning for a second attempt once the alarm has stopped, most systems have the facility of automatic re-set so that it will sound again almost immediately. Every system has a delay facility so that when it is set there is sufficient time for the householder to leave and enter the premises and immobilise the alarm before it activates.

Many people have a 'dummy' bellbox – as the outside housing for a sounder is called – fitted, and there is no doubt that it often acts as a deterrent. But burglars frequently attack this box first either to test if it is a 'dummy' or to try to prevent the alarm sounding. So the boxes – this is compulsory if they are to conform with the relevant British Standard for alarm systems – are designed to be tamperproof and with a separate battery power source so that an alarm will still sound even if there is interference, and will continue to sound even if the cable linking it to the control unit is cut.

Most of the do-it-yourself systems now being marketed require no wiring as they are either of the portable infra-red plug-in type and merely have to be sited in the right position, or, if they are of the type using pressure pads and door and window sensors, the wiring is minimal and manufacturers claim that such units may be installed by any householder who is handy with a screwdriver, power- or hand-drill, chisel and a pair of pliers. These systems have been criticised on two grounds; firstly, it is suggested that any burglar can walk into a D-I-Y shop or other stockist, buy one to find out exactly how it works and then just as easily devise a method of overcoming it; and, secondly, there is no expert control over the installation to ensure that everything is correctly done. And correct installation, correct siting and correct use are essential. Most evidence shows that carelessness is the cause of a great many false alarms from domestic systems.

Despite even the most sophisticated alarms, intrusions still take

Intruder alarm from Noise and Security Appliances which contains full microchip circuitry and can be used in a house, car or boat.

place, often due to the fact that entry is made by way of the parts of the premises not protected by the system. Comprehensive household packages, such as those supplied by the leading companies in the field – Chubb, Lander, Modern Alarms, AFA-Minerva, Securitas, and Songuard, to name but a few – give all-round protection. Most of them are two-zone systems so that either the upper or lower part of the home can be safeguarded while the other is not – which may be necessary when the premises are occupied – or the whole house can be protected when it is unoccupied.

With most portable systems, protection is given only to that part of the home which is most vulnerable, such as lower rear windows and the patio or french doors, because there is a limited choice of sensing methods. For example, door and window contacts cannot normally be used as this would entail some wiring and the portable would virtually become an installed system. There are exceptions, such as the Multi-guard Portable made by the Davco Instrumentation and Security Company. The feature of this is that it may be used independently as part of a more comprehensive system and can be linked to pressure mats, magnetic contacts and lights. The portable unit itself is an ultrasonic detector. Portables have become extremely popular, chiefly as a result of technological advances in recent years which have so much improved their performance, together with the fact that they require no wiring, can be purchased outright without the need for an annual maintenance fee, and have an aesthetic appearance.

Conventional burglar alarm systems have mostly been designed to detect a burglar moving about inside the premises after he has broken in. This means that damage has already been caused by the intrusion, and it gives the burglar the chance to collect valuables and then escape by either attacking the alarm first and neutralising it if possible or just by acting quickly enough to be in and out before the alarm response becomes effective. In his haste he may cause more damage, and there is the possibility of violence to an occupier who confronts him. Systems are therefore now being designed on the real crime prevention basis of detecting the intruder before he enters the premises and then raising an alarm calculated to scare him off. Probably the best known in this field is the EW2 Home Alarm designed by the Bedford firm of Notecalm Ltd, which provides protection for enclosed areas measuring up to 5,000 square feet. It operates on the principle of acoustic discrimination by 'listening' to the audible sounds in the environment it is protecting. By means of unique micro-processor-controlled circuitry it is programmed to 'recognise' the percussive noises made by an attempted forced entry, while avoiding giving false alarms by identifying noises associated with normal household activity and not reacting to them. The alarm is given by a powerful electronic siren,

Notecalm EW2 system which raises the alarm *before* the intruder enters the premises.

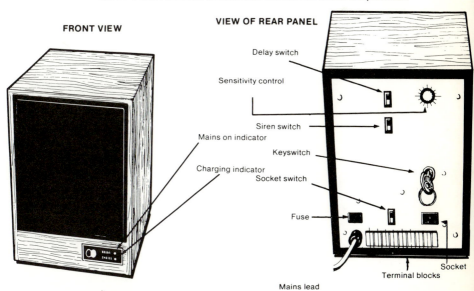

and if linked to one or more lights in the home can switch them on to give the effect of an awakened household if required. Although the unit is a plug-in portable it has the terminals and circuitry for wiring to a conventional loop system with contacts on doors and windows and pressure mats.

Guiness Security Systems also claim the elimination of most of the causes of false alarms with their professionally installed comprehensive system which works on the principle of detecting vibrations caused by intrusion attempts. The company claims the system is more dependable and more efficient than those relying on detection by sound, movement or change of temperature. The electronic brain of this system is an analyser which has two 'amplitude thresholds'. One of them is designed to operate the alarm if attempts are made to force, or to remove glass from, windows and doors. The 'memory' reacts to signals from tamperproof sensors on doors and windows, and is programmed to activate the alarm if a certain 'count' is exceeded. The second 'threshold' is designed to respond to a major attack such as a brick being thrown through the glass, in which case the memory is by-passed and an immediate alarm raised.

An indication of the increasing sophistication of modern home burglary alarm systems is also demonstrated by the Guiness installation. It has a facility which makes it possible to have the system in full operation even when the household is occupied – instead of having to switch it off as is the case with traditional alarm installations – so that a lone occupier can guard against surprise intrusion through the part of the premises not being used. Another novel feature is a switch which is fitted just inside the front door. One touch of this allows the door to be opened for the exit or entry of visitors while the alarm system is turned on, but if there is an attempt to force an entrance or the caller frightens the occupier in any way, a second touch will trigger the alarm.

The computerised 'Intercept' system.

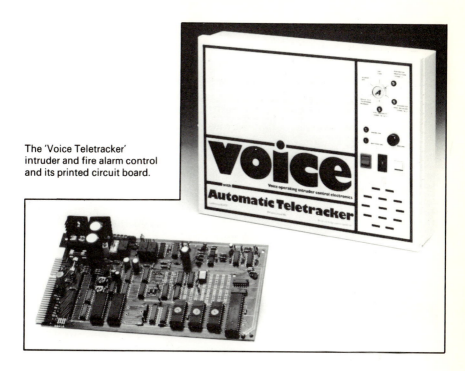

The 'Voice Teletracker'
intruder and fire alarm control
and its printed circuit board.

One very sophisticated system known as 'Intercept' is programmed to suit individual requirements and has been claimed as the world's finest computerised security system and the first of its kind in the UK designed for the home. Each household is given a secret code number so that the alarm cannot be altered or programmed until the householder has tapped in the correct number.

A revolutionary new system in which all the circuitry is contained in one printed circuit board with a single plug-in edge connector has just been launched by Kalami Ltd of Farnham Common, Berkshire. This is known as the 'Voice Teletracker', and has the ability to communicate instantly by the use of an in-built voice synthesiser. Using a human voice, it actually gives information and directions to the user, talks to the maintenance engineer giving him relevant data for fault-finding, and transmits an alarm over standard telephone lines by giving spoken details of the emergency if there is an alarm. And it shouts for either the police or fire brigade through an external sounder, giving the number of the house. By tapping in the number, the home-owner can programme the system to call him at the telephone number of his business or any other address he may be visiting.

Wireless radio-based alarm systems are now being installed in ever-increasing quantities, but a great many of them are not approved by the Home Office and their use is therefore illegal. A great deal of radio equipment for home systems is imported and is not only unapproved, but is unlicensable. In most cases it is not illegal to sell, buy or own the equipment – only to use it. But few, if any, devices using radio cause serious problems of interference because the transmitters produce only about one-hundredth of a watt and, owing to their small size, rarely have efficient antennas. Transmission time is less than a second, with repetition at hourly intervals. The official licensing agencies are not capable of policing or detecting this low-powered radio equipment and it is used with impunity.

It must be understood, however, that there is an officially allocated frequency for portable personal attack devices – perhaps more familiarly known as 'panic' devices. Some manufacturers have now produced equipment which meets the UK technical requirements and the frequency band 173.25 MHz has been allocated for use by approved fixed-type installations. It is really the transmitter which has to be approved, and all manufacturers and suppliers of this approved type declare it to their customers.

The potential domestic market for radio systems is huge for they eliminate the need for household disturbance such as running wires and taking up floorboards, and the cost of skilled labour. A typical local alarm with six window and door transmitters, a personal attack transmitter and two smoke detectors sending their signals to a central receiver control unit can be installed in less than an hour in many cases. Some experts have forecast that tens of thousands of radio alarms, many of them in domestic premises, will be installed in the next five years. But, they warn, this could lead to a large jump in false alarms because there are many problems of interference.

The growth of what are called central stations has done much to improve the efficiency of alarm systems. These stations are manned day and night by a highly trained staff and are operated by most, if not all, of the main manufacturers and many of the small outfits as well. Alarm systems linked to them are monitored and when an alarm is activated full details are automatically conveyed to the central station by means of a communicator. A visual display screen just like that of

Central station which monitors security and fire detection systems in addition to other services such as heating and ventilating, lift operation and commercial concerns.

a television set shows the origin and nature of the call. The operator checks it first wherever possible and calls the appropriate service – police or fire brigade – if the alert is genuine.

Many installed domestic alarm systems use a different signalling technique. An auto-dialling unit, for example, is activated by the control panel when an alarm is generated. It enables a pre-recorded telephone message for a 999 call to be transmitted over a British Telecom line to a manned exchange, where the operator relays the message to the police. This is the most common method for domestic alarms as the central station linkage mentioned above is comparatively expensive.

All good alarm systems, whether portable or installed, are self-monitoring and, when they are switched on, indicate either visually or audibly – or both – that they are in good working order. The National Supervisory Council for Intruder Alarms – known as the NSCIA – was established in 1971 with the support of the Home Office, police forces, insurance companies and the security industry, with the aim of improving the reliability and effectiveness of intruder alarms and to reduce the false alarm rate. It established, and maintains, a register of approved alarm installers who agree to install systems to the British Standard 4737. They are also expected to conform with the new NSCIA code of practice published in 1981 to supplement and to reinforce the British Standard. The work of approved installers is inspected for quality of workmanship and compliance with both the Standard and the code, and the inspectorate investigates technical complaints. Two years ago a training scheme was introduced for installers.

Any householder who wants security for his personal possessions can obtain technical advice on the type of security he wants prior to installation of the system. The NSCIA points out that there are many experienced and reliable installers, but warns that there are inevitably those who are out to make a quick financial killing and these should be avoided by choosing from the approved list. It advises choosing a system which conforms to the British Standard, selecting at least three companies from the approved list and, if there are operational problems which cannot be resolved with the company, asking the NSCIA to investigate on your behalf.

The difficulty facing the home-owner is that there is no statutory regulation governing installations and there are many reputable companies, especially the smaller ones, who do not register with the NSCIA. Another problem is that, at the time of writing, radio alarm systems cannot be inspected since, apart from the Home Office transmitter type approval, there are no British Standards or codes of practice which apply to them.

There is another organisation whose stated aims are the same as those of the NSCIA and that is the Independent Associated Alarm Installers Ltd, which represents a small number of installers. The IAAI describes itself as 'an organisation set up to benefit the consumer and one whose members are authorised to issue certificates of status confirming that their workmanship conforms to BS 4737 and/or the code of practice issued by the Association'.

4 A new kind of security

Over 20 per cent of households – that is, more than four million households in Britain – consist of people living on their own. Although single-person households are not necessarily elderly, in fact the majority of them are. Few people realise that over two-thirds of these single-person households consist of people aged 60 or over, which means 2.9 million of all households in this group. And four out of every five of these elderly people living alone are women.

This situation has brought about a new dimension in security needs. Although recent surveys have shown that elderly women are not the targets of criminals to the extent that many believe and that their fears of becoming victims are exaggerated, there continue to be savage attacks on the old and the infirm, both male and female. Quite apart from the possibility of criminal assault and burglary, however, it is the elderly who are particularly prone to accidents in the home. Over 60 per cent of such accidents involve a fall causing fractures of varying severity. Elderly women are more likely to fall than men: around 68 per cent of accidents in the home involving women between the ages of 65 and 74 and some 80 per cent of women aged over 75 years are of this nature.

Since more and more people are living longer, the risks to those living alone grows in proportion. Over and over again, an elderly or handicapped victim of either attack or accident has lain helpless for hours or days – and sometimes has died before anyone has realised that something was wrong. Alarm companies and systems designers have been researching this area in depth, always bearing in mind the determination of many of the elderly to retain their independence and dignity and to look after themselves rather than go into a home or any other institution, despite the risks involved.

A number of systems and devices have now been developed which combine the functions of intruder and fire alarms, and at the same time provide a 'cry for help' service comprehensive enough to meet any emergency. Unfortunately, most of the systems, although cheap in comparison with many ordinary household intruder alarm systems, are still too costly for many pensioners who live alone. But it has been suggested that the families of elderly people might consider pooling their resources and buying such a system as a Christmas or birthday present. It certainly would give peace of mind to the many people who are constantly anxious about elderly relatives living alone.

In most 'distress' situations involving elderly people, the victim suffers injuries, such as a broken hip, which prevent movement and prohibit access to a telephone or attracting attention in any other way. The victim may be lying helpless in any part of the home or even outside the premises. Any worthwhile system has to be of a type which can be used whatever the circumstances, and which incorporates some kind of automatic or regular monitoring to give a warning if the victim is rendered unconscious. The only really satisfactory system which incorporates these needs is the wireless radio-controlled variety. It has been pointed out in an earlier chapter that radio-controlled systems for domestic and personal protection are being installed or carried as personal alarms on an increasing scale. These personal alarms, for which a radio frequency is available, and the installed type which are now legal if they operate in the specified frequency band and are of an approved type, mean that future domestic alarm systems can combine all the functions just as those created specifically for the elderly do.

Until comparatively recently wireless systems held little appeal and almost all the equipment was imported because they were costly and had other disadvantages such as heavy battery current consumption. But the micro-chip has changed all that, and very complex systems based on miniature circuitry and easily controlled by hand-held transmitters are now becoming more and more popular.

Alert Systems, a London company which imports radio alarms made by the American manufacturer Linear Corporation, pioneered the supply of wireless systems in the UK. At first the equipment, which operated on frequencies used in the USA, was not licensable in this country although huge numbers were bought and are in common use. Alert were one of the first companies to enter the personal alarm market and, before any of the sophisticated wireless systems for the elderly were developed here, they were marketing a portable 'panic' radio device which allowed users to activate a siren to attract attention if they were in trouble. Their imported equipment is now of an approved type, however, and is being used in a variety of domestic systems. One of their well-known products, a remote-controlled cash carrying briefcase, may itself have household application in view of the increasing number of doorstep attacks. For instead of risking violence at the hands of a robber, the householder can allow the case to be grabbed, wait for the thief to run off and then activate an alarm and a smoke canister in the case by remote control.

At the time of writing, probably the most sophisticated system for the protection of the elderly is one introduced by a charity operating from East London. Called 'Carecall', it is the only one which is self-monitoring all the time, instead of at pre-set intervals, but an exciting prospect is the fact that it is being offered free to elderly people living alone in council accommodation and to the very needy in private housing, and can also be bought by private home-owners living alone.

Personal alarm system showing the 'panic' radio device.

The transmitter is a pendant-type necklace worn round the neck – or it may be carried in the pocket – and when the operating button on it is pressed, it transmits a signal automatically picked up by a revolutionary new type of telephone. Called the 'Care Phone' it was invented by 42-year-old Mr John Adams, an electronics expert who is operational head of Monitron Computer Monitoring Services Ltd, in London. The phone is just like an ordinary machine and is used as such, but it contains a receiver and a communicator. In its simplest form it is equipped with a panic button which, when pressed, will call the police silently. But when the button on the transmitter pendant is pressed it is picked up by the telephone automatically, even if the wearer is lying injured in the garden, and relayed to a special central station which is manned night and day. The operator who receives the call makes a check and takes the appropriate action by calling the required service.

Although the 'Care Phone' was originally designed for use by the elderly, it has been developed so that it can receive signals from a full radio burglar alarm system, evaluate them, sound an alarm siren and activate a strobe flashing light and simultaneously notify the police or fire brigade by means of a central station. Arrangements have been made for a donation to be paid to the charity Care Trust every time one of the new phones is sold. Already a company called Channel 1 Associates Ltd is using it as the control in one of a range of new burglar alarm systems specifically designed for the home and based on passive infra-red movement detectors of a special type designed to cut out false alarms. Channel 1 have launched a unique home security system called CARE which, at the time of writing, is the only genuine D-I-Y system which is completely wireless and yet offers the same security response and low false alarm rate demanded in high-risk security areas such as banks. It is radio controlled by a small hand-held switch and all alarm signals are automatically transmitted by the Care Phone over an ordinary telephone line to a central station where an operator first vets every call by phoning back or contacting a neighbour before alerting either police or ambulance. The basic D-I-Y kit consists of two British-made movement detectors, which are merely screwed or stuck into position, and a smoke detector.

Opposite page: Security briefcase: the alarm and smoke cannister within are operated by remote control.

Haley Help-Aid System showing the visual display unit (*back*), loudspeaker (*left*) and transmitter (*right*).

The Haley Help-Aid System produced by Haley Radio Security Ltd also works on the principle of a small cordless radio transmitter which is carried on the person, but in this system an audio-visual alarm is activated. The transmitter, slightly larger than a cigarette packet, may be carried in the hand, in a pocket or handbag, on a belt or round the neck. In an emergency the transmitter button is pressed and the range of at least 150 feet means that, even if an accident or attack leaves the wearer helpless in the garden, the alarm system still works. The alarm itself is a triangular box which is placed facing outwards in any suitable window and plugged into a power point.

When an alarm is triggered a light in the box continuously flashes the word 'HELP' so that a neighbour, passer-by or patrolling police officer can see it, while a pre-recorded message shouts for help through a loudspeaker plugged into the visual alarm. The system can be installed directly into the ordinary telephone line to either the telephone number of a relation or friend, or to the national Chubb 24-hour alarm service.

Lander Alarms have launched a similar personal security system called 'Lifelink' which uses a portable radio transmitter in the form of a pendant or a watch strap. This system works through an automatic communicator connected to an ordinary telephone socket in the home. What is described as a 'comfort' tone is emitted for the reassurance of the person in trouble, while a coded message is sent via the phone to staff at the Lander central station, who then summon help.

From Davis Security Communications Ltd comes another self-dialling communication unit designed for the elderly and infirm living alone. This is called 'Unidial' and, again, there is automatic dialling of a pre-set telephone number when the button on either the communications unit or a pendant-type transmitter is pressed. This company has a control station set up to monitor elderly people living in sheltered housing. Data is also transmitted so that even if the caller is unable to speak there is immediate identification.

Britannia Security Systems Ltd in Kent operate a radio distress call system linked to a computerised personal information bank which contains vital information about each subscriber. It is stored in their

Unidial self-dialling communication unit, with remote control pendant transmitter.

own central station and gives immediate access to details of health problems which might lead to a situation needing help, and to a list of friends and members of the family who may be contacted; and in the event of an emergency there is a special facility. If the subscriber suspects intruders or is alarmed about an unwelcome visitor, it needs only a press on the button of the transmitter and the use of a codeword to the operator who will answer the phone and Britannia will call the police immediately. This system also works through a communicator attached to the ordinary telephone. The transmitter may be either of the pendant type or worn on the wrist like a watch. To meet special circumstances – for example, if the user is confined to a wheelchair – Britannia also provide a wired system with one or more push buttons that can be fixed in easy-to-reach positions around the home, and which costs less than the wireless system. Britannia also produce a comprehensive computerised domestic security system.

A slightly different type of transmitter is used in the Vitalcall system recently introduced by Vital Communications (UK) Ltd. It is a pendant device but is operated by pulling the pendant itself, which is on a cord which may be worn round the neck or hung within reaching distance when in bed or in the bath. The transmitter may be worn either over or under the clothing, and Vital Communications claim that it has been designed in such a way that it can be operated by even the most arthritic fingers. As with the other systems, transmission is picked up by the telephone. Three selected telephone numbers are then automatically contacted in sequence, such as a relative, friend or a neighbour. When the call is answered a pre-recorded message is read requesting urgent help. If there is no reply from any of the selected numbers, a control centre is automatically alerted.

Many experts believe that in the next few years radio-controlled wireless domestic alarm systems will take over a big slice of the market as a result of the advantages they offer, such as ease and speed of installation and operation, the elimination of extensive wiring, the sophistication of hand-held transmitters which means both immediate controlled alarm activation, and the benefit of remote control arming and disarming of the system when outside the premises.

The systems specifically developed for the elderly, even if they are beyond the financial means of some pensioners living alone in owned or rented accommodation, are ideal for warden-managed complexes where a system may be geared to a single control in the complex. The option to purchase or rent is another benefit attached to most of the systems.

One area of security which has been almost completely neglected until recently is the situation of lone residents. These include not just geriatrics, but those of any age or sex, perhaps more particularly women, who live in flats and rented rooms – where a complete burglar alarm system would be both impracticable and unnecessary, and where even a small portable alarm unit based on traditional principles would pose problems – but who are nevertheless seeking some form of more comprehensive security than just locks and bolts. In view of the repeated instances of intruders gaining access to rooms where women are known to sleep alone, there is a very real fear among all such occupiers of becoming a victim.

At last one company, Pifco, has tried to meet this demand. Two new products they have designed give forms of protection which, at the time of writing, are unique. One is an electronic 'doorguard' burglar alarm which is easily fitted to the door of any house or flat and has a piercing 95-decibel alarm which sounds as soon as the door is opened. No wiring is required to install the battery-operated unit, which measures just over $6\frac{1}{2}$ inches by $2\frac{3}{4}$ inches by $1\frac{1}{4}$ inches. It will fit on any door regardless of whether it is left- or right-handed opening. The unit is disarmed when a personal three-digit code – which can be selected and set, and later changed if desired – is entered on the control panel. Once armed, a 'delay' feature allows seven seconds for the occupier to enter or leave the property without activating the alarm. Alternatively, the alarm can be set to sound immediately the door opens. There is also an instant 'panic' button which, when hit, sounds the alarm immediately, whether the door is open or closed. And there is a setting which causes a pleasant chime to sound instead of the siren, but still alerts the occupier, if that is preferable.

The second device from Pifco is an ingenious portable door alarm which could be invaluable for the protection of lone women such as nurses living in hostels, students living in rooms and occupiers of bedsits. It is lightweight, measures only 4 inches by 3 inches by $1\frac{1}{2}$ inches, and is easy to carry. Apart from use in the home, Pifco say that because it can be carried around it is invaluable when away from home as well – particularly when staying in hotel rooms or in a caravan. Operated by a nine-volt battery, it simply hangs on any door knob and its electronic circuitry sets up a detection field in and around the knob.

When that field is disturbed by a hand, key or lock-picking tool, the 95-decibel alarm is triggered. Obviously, the use of this alarm presents a false alarm hazard, and its use would have to be restricted if genuine callers are expected.

Some people are not prepared to introduce burglar alarm systems into their home at all, especially if they are the 'stay-at-home' type who rarely leave their property unoccupied except when going on holiday. For such people, there is a service which will provide them with 'occupation' security. A firm called Homesitters Ltd, which has branches throughout the UK, supplies 'homesitters' who will take up residence in anyone's property while they are away, whether it is for only a few days or several weeks. This service gives protection against burglars, vandals and squatters, eliminates the chore of 'shutting up the house', and there is the added bonus of saving the cost and worry of putting pets in kennels or other temporary accommodation. The company has more than 300 homesitters on its books, many of them retired civil servants, professional people and police officers, all of whom are mature and responsible citizens who have been carefully vetted.

One of the areas where security has always been very difficult to plan on a cost-effective scale has been the control of access by un-authorised persons to domestic high-rise premises. Door and phone entry systems have been discussed earlier, but these are very expensive and do not necessarily prevent attacks on residents using lifts. And, of course, there are many high-rise blocks where there are no entry control safeguards and where the installation of a conventional-type system would be prohibitively expensive or just not practicable. There is, however, a sophisticated but simple-to-install package now available from the Surrey firm of First Inertial Systems. It regulates access by a micro-processor-controlled system which entails no additional cabling or the prohibitively high costs usually associated with access control equipment normally used only for commercial and industrial purposes.

The First Inertial systems are very compact, self-contained and designed for installation in the lift car itself. They are operated by individually programmed nylon keys which the lift user places in a reader when he enters the lift. He then selects the floor button on the main console and, provided his key programme is authorised for that area, the lift will function. The system allows for up to 1,000 individually programmed keys which are completely anonymous and which can be fixed to key rings. If lost, they are deleted from the system. The system operates up to 32 floors, and eight time zones which can be individually programmed to permit tenants to select the times they require lift security to be in force. The sequence is controlled by a micro-processor with a real time clock, which allows changes in the programme to be made without any inconvenience.

Barkway Electronics have designed for smaller blocks of flats a door entry control system named 'Minigard' which is claimed to be extremely resistant to vandals. The front plate of the Minigard is electronically 'latched' on to a recessed back box so that no visible fixings are in a position to be tampered with. It is claimed that even if someone hits the keys with a hammer they will have great difficulty in disabling the system, and the speaker grille has been specially protected to prevent anything being pushed inside it. Minigard is suitable for between two and 18 apartments or areas requiring two-way speech between the callers at the entrance and the occupiers, who can

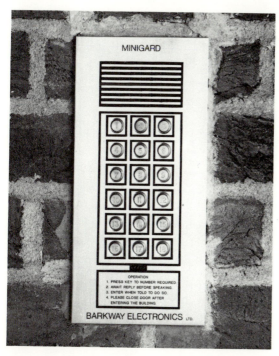

Minigard door entry control system for small blocks of flats.

then allow visitors to enter using a remote-control door switch. Bark-way operate a warden call system for sheltered housing and the system can be used in conjunction with it. The company also produces a system similar to the Minigard for larger blocks of flats, which offers a high level of speech quality.

Interference caused by car ignitions, CBs, heavy electrical equipment and other radio frequency sources has always caused problems for radio alarm systems by triggering off false alarms. Thrust Technology Ltd claim to have solved the problem by designing a new alarm system that transmits a complex digital code pattern which must be received clearly four times by the receiver before it will trigger an alarm. The company maintains that, once the link is established, severe interference will be tolerated, thus enabling the system to operate in conditions of strong radio noise and attenuation. Each transmitter and receiver has a ten-position switch to permit the selection of any of a total of 1,024 different binary patterns. The receiver

Hand-held interference-proof radio alarm which transmits a signal as long as the button is depressed.

will activate only if it receives a code pattern which has been programmed into it, enabling several systems to be used in close prox-
imity without any danger of false triggering.

There are three types of self-contained transmitters. One is housed
in a case that can be carried in a pocket or handbag or on a belt, and
transmits as long as the button is depressed. The second is identical
except that it will maintain transmission for a pre-set period. The
third is a lightweight pendant type worn around the neck on a lanyard
and is activated by being pulled downwards. This last device is designed primarily for use by the elderly and handicapped living either
alone or in sheltered housing. One very interesting point about these
devices is that, although the first two are powered by nine-volt alkaline
batteries, the third, because weight was a constraint, posed a problem.
So the Duracell company was asked to advise and came up with small
lithium manganese dioxide batteries, three of which are used in this
transmitter to provide a nine-volt battery pack with a shelf life of
between five and ten years.

Of course, whatever the future holds, it is still a fact that most
domestic burglar alarm systems do not yet have the facility of an
automatic response – they depend usually on the response of people
who hear the alarm siren sound. It is interesting to note that one
company, which produces the Home-Minder do-it-yourself system
aimed at the mass market, includes in its pack 'Good Neighbour'
cards to advise adjoining householders that an alarm system has been
installed and telling them what to do to help if it sounds.

5 Attack-resistant materials

The use of attack-resistant materials – widely used in public buildings and commercial and industrial premises – has only a limited value in the home and it is rarely necessary for the ordinary householder or flat dweller to even consider them, except in special circumstances. It does, of course, depend upon the kind of risk involved and the value of the contents of a home. Wealthier householders and luxury flat dwellers, or those occupying homes in remote country districts, especially if they have valuable possessions, may consider that the extra security provided may well be worth the extra expense. In some cases, of course, a survey by the insurance company may be followed by advice to incorporate such safeguards to prevent premiums being greatly increased or even a policy being refused.

The use of barrier-type security is usually expensive because of the nature of the materials used, the fact that very often they must be tailor-made for the premises, and because structural alterations may be needed. It is obviously useless to have security glazing set in woodwork which is rotten, or to have bars and grilles attached to weak walls and flimsy frames.

Grilles, shutters and bars provide much better protection to windows than locks. But they are obviously much more costly and, for the ordinary home-owner, their appearance alone, suggesting a fortress rather than a home, usually prohibits their use. I believe they may also pose a security risk which outweighs their security value if they are installed in ordinary domestic premises without a very special reason. Any thief might well decide that, if such protection is thought necessary, the contents must be well worth stealing. The amateur might be deterred when faced with such obstacles, but the professional burglar could be encouraged to take the risk of attack in the knowledge that, if successful, his reward might be that much greater.

Although grilles and bars are traditionally made of bright mild steel which gives much greater resistance to sawing and cutting than iron or black steel, there is a very wide choice available today. Aluminium and even a tough PVC material which is covered with anti-corrosive paint gives an attractive appearance to both grilles and shutters, which are designed so that when they are closed they present to all intents and purposes a single and compact surface. A country cottage effect may be achieved by the use of fixed flat bar or square bar grilles. But few people like this type of grille and the most popular ones are either the folding or sliding variety which are fitted inside the window and which give excellent security when closed and an uninterrupted view when open.

There are a number of special situations where the use of bars or grilles may be considered necessary to protect the windows or other glass areas of ordinary homes. One example is the decorative type of metal grille used on the inside of glazed front doors and porch doors to prevent an intruder cutting or breaking the glass and putting his hand through the hole to manipulate the lock.

There are some homes with vulnerable windows at ground level, especially in older dwellings in high crime rate areas of inner cities, for instance, which not only present the opportunist thief with an easy means of entry but are also an invitation to vandals. Few people realise, either, that many burglars do not break or cut glass, but if a window is not too large, and especially if the putty is old and in poor, cracked condition, simply remove it with a knife or some other implement and lift out the whole pane. Strong bars, shutters or grilles will give very good protection indeed. It is possible in some hardware shops to buy D-I-Y burglar bar kits, but these must be very carefully installed or the bars may easily be wrenched out of their wall fixing points.

Another device which can be fitted easily by a home-owner is the Shaw Lock-Bar for the security of top-hung night vent windows through which burglars often reach down to get at the casement fasteners of adjacent windows. When fitted, the bar cannot be disconnected unless the window is firmly closed and even then it may be released only from the inside.

One manufacturer, the Bolton Gate Co Ltd, has recently produced a new folding security grille for domestic application. Constructed on the lattice and picket system which gives both strength and flexibility, it is designed to fit over windows of all sizes. The screen is hung on

Folding security grille in the secure close position (*left*), and completely hidden by curtains when fully bunched (*right*).

rollers from a top rack fastened to the lintel, which can be completely hidden behind a curtain rail or pelmet. The bottom edge of the screen is located in a channel guide which can be removed when not in use. A feature of the construction is that the grille folds away to a very small area when not in use. There is also a version of this grille available for fitting inside patio doors – a concept which may well appeal to a great many home-owners. There are many break-ins recorded by the police in which burglars have used suction pads to lift patio doors off their runners, silently carry them a few yards and just stand them up against the wall of the house, thus giving the intruder quick entry and exit.

The sight of a strong metal grille inside patio doors might well deter the intruder from any attempt at entry, in the same way that the sight of a metal grille inside a window might. It is a concept of deterrence which could certainly be taken into consideration at the design, planning and construction stages of housing. Burglar alarms are devices for detecting intrusion and attempted entry and then summoning help, but such automatic detection should never be viewed as an alternative to effective physical security. Locks and bolts form a basic part of that security, but the use of barriers which look substantial, may take a long time to penetrate, and may be penetrated only to the accompaniment of a lot of noise, could well be a much more significant and worthwhile investment.

Modern homes almost always include a large number of glazed areas, some of them in single panes of glass, and this has always posed a security problem because of the brittle nature of glass and the ease with which it can be broken or cut. After all, in remote areas and high-rise dwellings where a great many flats may be unoccupied at the same time, a blow from a brick is sufficient to gain entry and is relatively safe because there are no neighbours at home to hear the noise.

The answer could well be the increased use of laminated glass in the home. This is a sandwich construction of annealed glass and polyvinyl butyral, with the glass forming the outer layer, which provides considerably greater resistance to any attack of the sort expected from a burglar by such means as a brick or hammer blows. It is made in various thicknesses for different purposes, is now used to give security in the glazing of many types of patio door and has been recommended for use in secondary double glazing.

A revised code of practice introduced by the British Standards Institute lays down new requirements about the use of safety glass in both single- and double-glazed doors and side panels. The responsibility for specifying the correct type of glass rests with architects and glass specifiers but an independent research carried out some time ago showed the regulations were little understood. A great deal has been done to try to popularise the use of laminated safety glass by companies such as Triplex and Alcan Safety Glass Ltd. Alcan, who are one of the biggest manufacturers of this material in the country, launched a countrywide campaign in 1982 which highlighted how little was understood about safety glass.

Perhaps the most widely used material in place of glass for secondary glazing of external windows is polycarbonate sheeting: a very tough plastic material indeed which is virtually unbreakable and yet may be cut to size quite easily. It gives a high degree of protection which may be varied according to the thickness of the sheet.

In addition to the security factor these glass substitutes are also very important in the home for safety reasons and are therefore a vital factor in personal protection. In households where there are both young children and internal glass doors, the risk of serious accidents is always a major hazard. Every year there are appalling incidents involving not only youngsters and the elderly but people of all ages who trip or are taken ill and crash through a glass door. Using laminated glass or polycarbonate sheets gives vital protection.

Where security of a particularly high standard is required by a home-owner, burglar-resistant doors – which have already been mentioned briefly – may provide the answer. A range of doors distributed by the Burt Boulton Architectural Co includes security doors made of very thick and attack-resistant plywood with a manganese steel plate in the heart of the door and formed into a U-section around the locks. Special anti-burglar hinges and locks are available too.

Another company, Magnet and Southerns Ltd, has also evolved a range of security doors for domestic use. The company offers three types of door in this range, each one providing a different degree of security. One of them is a square timber-frame door faced with galvanised steel panels which is injected with dense polyurethane foam and is supplied pre-hung in an aluminium frame. The toughest in the range is a steel door with steel reinforcing bars placed at five-inch intervals and covered in decorative wood-grained vinyl. This one is supplied in a steel frame. In addition to the security aspect, the doors have fire resistance ratings and conserve energy because of their insulating core. This latter feature is obviously an important factor to consider when deciding on the cost-effectiveness of such protection. All three doors in the range include a four-directional locking device which incorporates a handle-operated latch. They are deadlocks and, instead of being surface-mounted, are morticed into the body of the door itself. The hardened steel bolts on each side of the door shoot into boxed recesses which protect the ends of the bolt from any attempt to force them back. It is claimed that the lock cylinder gives unique security by its special design, with over 100 million differs.

Crittall Windows Ltd have also launched a steel security door which can be used for the home. It is actually a complete door set and includes a frame. Intended primarily as a replacement for ordinary doors, it can very easily and quickly be fitted into an existing wooden frame by means of concealed screws, after the original door leaf has been removed. Both door and frame are made of hot-rolled mild steel which has been hot-dip galvanised, and is therefore claimed to be 100 per cent rustproof. The door's stiffened steel panels are of heavy gauge. Fittings include a five-pin cylinder lock, a letter flap, a safety chain and a door viewer.

One relatively cheap and simple method of protection for glass in the home is the application of shatterproof film which will stand up to attack by a brick or metal bar and which is a very efficient safety measure to guard children from flying fragments and jagged edges if violent collision with the glass does result in breakage. Shatterproof film is a comparatively recent innovation. It first came into use in 1970 and became widely used as a result of the terrorist bombings on the mainland of Britain between 1972 and 1974 when it was necessary to do something to protect shop staff and the public against the danger of flying glass from blasted windows.

Since then, very exhaustive tests have been carried out by the 3M company who are probably the main producers of the material, and the film they market has proved highly successful as a protective barrier. Research and development have made the film of today much more sophisticated. One of the latest types of film available is called Profilon. It is one of the products of the Saracen Safes and Security Co, and consists of a transparent polyester film applied to glass with a special adhesive. It is claimed that glass protected with this can withstand a number of attacks and will not disintegrate into fragments. There is no doubt, however, that the use of these protective films is restricted in its application to residential property under normal conditions – it is more usually to be found on the windows of shops and similar premises.

Many householders already have one type of window barrier without realising it and those who would not dream of having a screen or shutters fitted might well consider it. This is the ordinary venetian blind made of metal strips which, when in the closed position, gives quite good protection.

6 Protecting portable property within the home

Probably the most difficult and the least researched area in which to assess the possibility and effectiveness of security is that of portable valuables within the home. Until they become victims, few people realise that even bulky and heavy items such as television sets are stolen regularly. One of the most attractive pieces of equipment for the modern thief is the video recorder now to be found in many homes. Computers, many of them ranging in value from a few hundred to thousands of pounds, are being added in increasing numbers to the costly items presented to burglars on a plate when there is inadequate security. And many home-owners and tenants do not realise, or decide to ignore, the fact that most household policies require that the insurers be told if such expensive items are added to the contents of the home. If this means a big increase in the contents value, a special premium may be needed or at least an increase in the existing one. If this is not done and there is a burglary in which the items are stolen, the company may not pay out at all or may agree to reimburse only a proportion of the loss.

Such equipment as hi-fis and expensive cameras are easily transportable, especially if the thieves have access to a waiting car – which they will probably have stolen as well, just for the purpose of crime. Although sophisticated methods, such as electronic tag systems, for the protection of these items are available from retail shops and big stores, they have little application to the domestic scene except in one or two instances involving small units giving single item protection.

As a result of the demand for the protection of videos from theft, a new invention has recently been marketed – the Videoalert, which can, in fact, be used for any other portable equipment too. It is an alarm which is smaller than a couple of cassettes, weighs only 13 ounces and is attached to the side of the item by means of 'supergrip' pads. It is switched on with a security key. If a thief tries to remove the item, it triggers a piercing 98-decibel scream, and the two batteries which operate it ensure that it continues to scream for up to eight hours. It re-arms automatically every two minutes after a 30-second pause unless it is switched off or the movement is discontinued.

Another completely new alarm for the protection of videos, but of equal value for the security of other items of electrical equipment including computers, is now being manufactured by Rothtron Electronics of Desborough. It is coupled into the mains lead. As soon as the equipment is unplugged, or if the lead is cut, the alarm is sounded and an internal battery ensures that the alarm continues to sound until the battery goes flat. The manufacturers claim that, once the alarm is set, it is impossible to disconnect the video from the mains without triggering it.

Perhaps the first thing to remember about portable property, however, is never to make it obvious that the home contains such a valuable item as a video recorder by positioning it in or near a window where it may be seen easily from the outside and attract the attention of a potential thief. Apart from this obvious precaution, any attempt to create some sort of physical security for each item may in itself lead to severe damage being caused in the home. It has been pointed out already that, once a burglar has gained access, he has plenty of time to inspect and collect the items he chooses to steal. And as the premises are usually unoccupied, it is unlikely there will be any risk of noise being heard. In these circumstances an ordinary cupboard door which is locked may, and probably will, be broken open. If an item is fixed to furniture by a chain or some other form of bond, the burglar will wrench or cut it free, smashing the furniture in the process.

Although several devices have been produced specifically for the protection of single items in the home, there is the possibility that the use of something designed for retail premises might, if it acted as a loud alarm, or if it meant that very considerable force and time would be needed to remove it, deter at least the opportunist burglar who is in the majority and who likes to get in and out as quickly as possible. Now such a device is available. Originally designed for video retailers in East Anglia who, like such shopkeepers everywhere, were losing the recorders on an increasing scale to thieves who simply picked them up and walked off at busy times, or when backs were turned, this very successful device is known as the Displayguard. It deters the thief by emitting a piercing scream when the object on display is lifted. It is made in smooth white glass fibre – other colours are available to order

– and weighs only one and a half pounds. Displayguard has smooth sloping sides which produce a display area much smaller than the base, making it difficult to pick up. And if the item on the stand is lifted, it triggers the alarm which can be heard as far as 100 metres away. It can display any object weighing between eight ounces and one hundredweight. Tests have shown that this stand will go on screaming for one and a half hours before the nine-volt battery which powers it gives out. The battery itself is tucked away on the underside so that it cannot be tampered with. It is possible that this device might well prevent the theft of a video and other electronic items in the home. It is cheap, needs no wiring, does not have to be fixed to furniture and can be moved around. It is worth a try!

Available from Floral Wire Products is the video clamp, which is a low-cost unit constructed of steel. It comprises a rectangular base with fixing points for permanent siting. The base frame is adjustable to accommodate any current model of video recorder, with four welded vertical clamps to secure the machine. Once adjusted and secured, the unit allows normal operation of the recorder. It is, of course, intended for the retailer and for such organisations as the BBC which has bought a number to safeguard its machines from theft. In the home, in my opinion, it might also be effective because it is a very strong device and it would give a burglar a great deal of trouble to remove. But the fact that it has to be screwed to the furniture is not a point in its favour – although a competent D-I-Y householder could make a table or stand specifically for the purpose of taking the clamp. In any case, at the time of writing, Floral Products are already considering the possibility of arranging the manufacture of a complete home unit which would be free-standing and would eliminate the need to screw the stand to existing furniture.

There is, however, one novel approach to this problem which can give a high degree of security not just to the video recorder, but to a wide range of other valuable equipment to be found in the modern home, including the television set. It is an 'Anchor-Pad' system which was pioneered originally for the protection of costly computers by the Hertfordshire firm of Data Design Techniques Ltd. The theft of office equipment has become big business since the introduction of sophisticated electronic machines such as computers, word processors, typewriters and visual display units. They all present attractive targets to the burglar because they are easily transported and easily

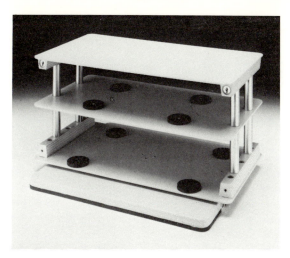

'Anchor Pad' system which offers a high degree of security to domestic items such as videos and television sets.

disposed of to an army of eager buyers. But although the 'Anchor-Pad' system was intended to save business and office losses, the growth of the home office containing similar electronic equipment has brought the same security headache to a growing number of private home-owners. And apart from the office in the home, many people are buying expensive micro-computers for their own use and pleasure.

Different from almost any other locking device, the 'Anchor-Pad' has two very important features for the benefit of the home user – it does not damage furniture and it can be removed if necessary. The unit designed for the protection of one particular range of computers is obviously of use only to those owning one of those machines. But the general purpose unit – which may be used to provide a high degree of protection for a wide variety of equipment ranging from any make of computer to the television set, video and electric typewriter – offers the householder, and particularly those with a home office, a solution to their security worries.

The 'Anchor-Pad' uses an adhesive base which is stuck to the desk or table surface and grips with a minimum force of 50 pounds per square inch. Properly installed, the company claims that it takes a force of about 5,000 pounds to prise the pad free. A 'mating' anchor-casting is bolted to the main chassis of the piece of equipment and the two are 'married' by means of heavy steel rods inserted by a special

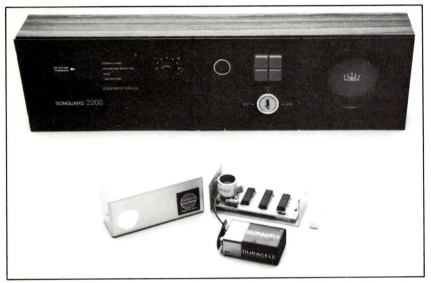

'Bug' with detector unit (*back*). When the small magnet in the 'bug' is disturbed, the device transmits an ultrasonic signal to the main module which in turn triggers the alarm.

tool. As a further precaution, numbered key locks are inserted into the casting to prevent any removal of the rods. Similar pieces of equipment are interchangeable between locations. If it is necessary to remove the 'Anchor-Pad' it is done by means of a special liquid.

Whether there is any merit in fitting alarms for the protection of specific items is debatable but there are reasons which might dictate a secondary line of defence: for instance, if a burglar has by-passed the home alarm system. And, of course, some home-owners may prefer to protect particular items or small areas where they keep valuable equipment, rather than the whole of their premises. Many of the alarm manufacturers produce small stand-alone battery-operated units which may be sited anywhere and which will protect either a single room or a single item. These can be placed among books and are so unobtrusive that there is no indication of their real nature. One company, Peak Technologies, offers a range of sensors to suit a single room – one of their sytems is called 'Roomwatch'. Songuard Ltd produce 'Bugs' which are battery-operated contacts fixed in place with adhesive tape which may be used to protect single cupboards.

One new development for the protection of personal property and cash is the 'Alarm Box' from Quirefive Ltd. This is made from a tough engineering-grade resin and has a high-security key lock with an alarm which, once activated by even the tiniest movement, emits a piercing scream. It is available from many stores.

There has always been a great demand for home safes and a very wide choice is available from many different manufacturers. As with locks, of course, just any old safe is by no means secure. Advice should always be sought from an expert before a particular safe is decided upon, and only those which conform to recommended standards of strength to resist attack from all the usual methods used by safebreakers should be considered. Good safes are not cheap, but a high price does not necessarily ensure qualities which will stand up to all the hazards which might be expected. Fire, for instance, is a very important consideration. It is no good buying a safe to protect important documents or banknotes if they are likely to be charred to ashes should the premises catch fire. With the increasing use in the home of technology which uses computer media such as audio cassettes, magnetic tape and floppy discs, the question of their storage in conditions which control temperature changes and humidity is very important and both safes or cabinets are obtainable which will meet all requirements.

If all that is required is a method of securing ordinary filing cabinets which can usually be opened with very little effort but whose contents need protection from possible vandalism, Guardall Ltd have developed an ingenious little alarm called the 'Chubb President'. It is

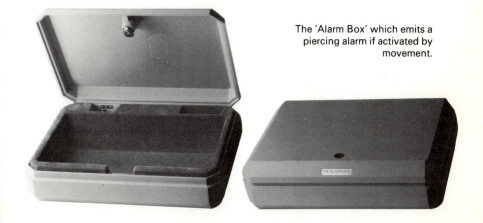

The 'Alarm Box' which emits a piercing alarm if activated by movement.

Above: Underfloor safe showing the unique ring bolt which operates all round the six-inch diameter circular door.

Left: Wall safe recessed into a wall.

only a few inches square, and can be fitted by any handyman to filing cabinet drawers or to any other similar container of valuable documents. Turning the key on the existing lock is all that is necessary to activate the system and, if anyone tries to open the cabinet without disengaging the alarm, a piercing siren sounds. Anyone authorised to open the cabinet has only to set a combination code and then turn the key in the normal way. A green light shows when it is safe to open up.

Most popular home safes are those which are either recessed into a wall, or in the floor. There are many different models designed to resist attack by drill, brute force, lock manipulation and explosives. Some companies, such as Secure Safes (Coventry) Ltd, supply a range of safes specially designed for the home which feature solid steel doors and anti-explosive bolts. Called the 'Householder', this range is claimed to be strong enough to withstand sledgehammer attack and it has a seven-lever lock with anti-pick features.

Chubb and Sons Lock and Safe Co Ltd manufacture a type of safe for domestic purposes incorporating 'long throw' moving bolts on the front, top and bottom edges of the door which are supported by deeply engaged fixed bolts on the rear edge, giving very great strength.

Hamber and Whiskin Engineering of Essex have developed two revolutionary new security ideas in their range of underfloor safes for the domestic user. One type has a unique ring-shaped bolt which operates all round the six-inch diameter circular door instead of using the usual shaft-type bolts, and this ring bolt gives greatly increased resistance to explosive force. The inventor and manufacturer, precision engineer Henry Bamber, has also incorporated in the safe a secondary mechanism to prevent the door being opened if the main lock is dislodged by explosives or any other means. The other new underfloor safe from this company is the Hamber 'Minder' which has vertical bolts that revolve into position instead of sliding horizontally, and has been specially designed to meet the demand for a low-cost, high-security safe for the home.

Experience has shown that there is a demand for a special kind of safe to fill the needs of home-owners who want a unit which gives greater strength and security than is provided by a conventional wall safe, who have neither the room nor the means to acquire a free-standing safe, or who live either on an upper level, in a mobile home – or even on a luxury yacht – and are therefore unable to have an underfloor unit installed. The Newcastle-under-Lyme company of Churchill Lock and Safe Co Ltd has introduced for this purpose a solid steel safe which does not need to be set in reinforced concrete as do ordinary wall and underfloor safes, but has a base fixing with four alternative fixing points.

If the items need protection against the risk of fire only – as in the case of valuable documents, for example – a vast range of fire-resistant cabinets, many of them with a high security rating, is also available and these save both money and installation disturbance.

It must be remembered, however, that the official advice from the police, Home Office and the British Insurance Association is never to leave cash or valuables of any kind lying around the home, and this is particularly important when you go away for even a few days. Some people seem to think that there is little risk if their absence is only for

Using an ultra-violet marker pen to inscribe the postcode and house number on valuable household items.

a limited period. The burglar needs only a limited period – a few minutes will often do. Burglars know from experience all the likely hiding places and, make no mistake, they will find them. The police advise that you should keep your valuables such as jewellery and cash in a bank or safe deposit box. The growth of the modern safe deposit centres using electronically controlled premises has meant that it is now comparatively cheap to rent a box, and many people are regarding such centres as an extension of the home as far as the protection of their valuables is concerned.

Property marking as a means of protection and identification is now being promoted by the police, the Home Office, the British Insurance Association and every law-enforcing organisation on a national scale. It is claimed that this may be one of the most effective ways of beating the burglar. It is quick, easy, and it is a do-it-yourself operation. It simply means putting your postcode and your house or flat number on every item in the home. The police point out that if property can easily be identified it acts as a very effective deterrent because thieves rarely take property which can later be traced back to them without difficulty. Secondly, if property is stolen it not only makes it easier to trace, but means there is a positive identification if it is recovered; sometimes that is not possible without this kind of marking because there are all sorts of items which otherwise look identical.

It is the postcode which is the vital link. Each code is just a simplified address and each part of the code focuses on a progressively smaller geographical area. Few people realise that each complete postcode identifies an average of 15 adjacent addresses. If you add to that the house or flat number – or the first two letters of the house name – you have a unique reference code pinpointing each address. More than a little useful if a thief is stopped in the street a few blocks away from his victim's home and still in possession of the marked loot!

There are three methods of marking. One is engraving and this is done with a fine drill or other sharpened tool. It does not matter whether you use a stencil or a template – or just do it freehand. The second method uses an ultra-violet pen giving an 'invisible' marking which can be read only under ultra-violet light. The third method is by punching and this is normally done using a hammer and a set of punches bearing the marking information.

Different types of property obviously need different marking methods. Invisible ink marking is most suitable for antiques and all

Using an electric engraver to stencil the number onto valuable property. The stencil is then removed to reveal a permanent record of ownership.

articles which might be devalued by being obviously marked, or damaged, and is eminently suitable for very delicate items although it can, in fact, be used safely on almost everything. But it is not as permanent as the marks made by other methods and may need renewing from time to time. The ultra-violet markers, which are manufactured by companies such as Berol and Volumatic, are available from most D-I-Y stores and stationers and cost only around £2.

Engraving is the most suitable method for items such as jewellery, cameras, television sets, video recorders and silver plate. For those with good eyesight and a steady hand, any expense involved in buying an engraving tool or kit may be saved by scratching the marking on the item. In the United States a computerised engraving system which has been in operation for several years has proved very successful in crime prevention – a survey of 150,000 households in which the system had been used experienced 11 per cent fewer burglaries than homes not similarly protected. It has now been launched in this country by the British firm of Guardmark. Subscribers to the system are given an exclusive code number which, together with their name and address, is entered into the company's central computer data bank. With the aid of a special electric engraving tool and stencil, the computer code number can then be etched onto television sets, video recorders, hi-fi equipment and other valuable items, leaving a permanent record of ownership. Once applied, the code cannot be removed without the object being damaged and, to deter thieves by letting them know that the items are so marked, Guardmark supply adhesive window and doors shields telling would-be intruders that the marking has been carried out.

Punching or stamping should be reserved for heavy items only, such as lawnmowers and bicycles – but be warned that any item made of aluminium should not be marked in this way as it is a soft metal and easily damaged.

It is important to remember that where you mark your property is important – particularly if you are using the engraving method. Since you will probably prefer the mark to remain out of sight you'll obviously choose somewhere behind or underneath the article. It is just as important to remember to select a surface that cannot be removed without spoiling the basic appearance or performance of the item, in order that a thief could gain nothing by so doing.

Left: 'Superswitch' model which replaces the ordinary lightswitch and automatically turns overhead lights on and off at pre-set times. *Right*: 'Superswitch' plug-in model which can be used to control table lamps and other electric appliances for 24 hours.

Cycle stealing is one of the most common forms of theft and machines should always be marked – in more than one place if possible.
You can protect your car by using a special window etching kit and
this can also be used on glassware such as decanters. Nowadays everything which is remotely useful or attractive is re-saleable, and therefore a target for thieves, so do not forget that the marking scheme
embraces everything you own from pictures and ornaments to furniture and the washing machine.

The police say that stickers strategically placed anywhere a burglar
might force an entry provide a very effective deterrent. Most police
forces will supply these adhesive labels bearing the legend 'Marked
Property' to the public – in London they are available free from any
police station.

In addition to marking property, a note should be kept of the serial
numbers stamped on many items such as bicycles and all electrical
and electronic equipment. The police also suggest you keep photographs and descriptions of the property you own, particularly of items
which cannot be marked. If possible, the photographs should be in
colour and special distinguishing marks such as crests, initials, hall-
marks or any other identifying matter should be shown.

Security lighting both for the interior of the home and to cover the
exterior in an emergency is an economical and very effective procedure
that is very often completely neglected. When the home is unoccupied
and left in complete darkness, its vulnerability is advertised to any
potential burglar. By always leaving lights on somewhere, the appearance of occupation helps to keep intruders away. But do not leave a
light on in the hall each time as this may give the game away. Leave
more than one light on in different parts of the house and vary the
rooms involved. The 'Superswitch' is ideal for this purpose. One type
is built into a security light switch. It is easily fitted in minutes,
replaces the ordinary switch and automatically turns lights on and off
at pre-set times. The second type simply plugs in and can be used to
control table lamps and other electrical equipment. Exterior lights
fitted in strategic positions can be used to floodlight the outside of the
house and the front and back gardens if anything suspicious is heard
during the hours of darkness. These lights may also be linked to the
burglar alarm system so that if it is activated they are automatically
switched on.

7 How to protect the person

So-called self-defence training is usually concerned with the ability of women to protect themselves against attack. It is the woman alone in her home who is more likely to become a victim of violence at the hands of an intruder and, as has been mentioned in an earlier chapter, rape has been perpetrated in many cases by a burglar or burglars who enter residential premises. Such attacks have led to the setting up of classes specifically aimed at teaching women how to defend themselves. No doubt there are circumstances in which women, especially if they are young and well-trained, may be able to disable a male attacker and send him packing, but there are very mixed feelings about the value of this approach. Many security experts believe it to be a dangerous exercise which not only has no real value but is more likely to put a victim in even greater danger.

Official advice puts the emphasis on teaching everyone, and particularly women, how to avoid as far as possible getting into any situation which makes them vulnerable. General advice by the police to women is to try to cultivate in themselves an awareness of what is going on around them. It is often possible to sense something before it actually happens, and recognition of something suspicious may lead to avoiding action being taken in time to prevent danger.

However, if despite such measures you find yourself confronted by someone, the police advise that you keep calm, for there is plenty of evidence to show that, by reacting violently or by screaming, you may frighten or goad the potential assailant into using violence which might otherwise be avoided or mitigated. Anyone confronted with a deadly weapon or obvious physical threat is advised not to resist as survival is the most important consideration.

Both reaction and the possibility of escape are often governed by the circumstances in which the aggression takes place, and this alone makes general advice difficult. To be confronted in the confines of your own home may mean that you are trapped; but if it happens in the open, it may be possible to escape and run and to scream if there are people around who might come to your aid.

One well-known security organisation, Sterling Guards Ltd, realising that there was a demand for advice and help, runs a very popular 'awareness presentation' entitled 'Preventing Assaults on Women'. The presentation, which is both audio and visual, draws attention to the fact that many people rely too heavily on hastily learnt self-defence techniques, which are forgotten or rendered virtually useless in a real situation. The company stresses that the main consideration is to prevent the situation in the first place. But, advises the company, if you are caught out, forget the knee in the groin or trying to kick - it puts you on one foot. Use your head to butt and your elbows to strike backwards. Lower your balance and try to keep on your feet.

To be of any use at all, self-defence techniques need regular practice and in most cases this is either not possible or is neglected. Even if the victim has the ability to disable an attacker, there is usually no time to do so because of the elements of surprise and panic. It is the panic and fear which the aggressor creates which often renders his victim both speechless and helpless. But women who are capable of remaining calm and who are trained in some form of self-defence may stand a better chance, and it may be helpful to attend a course of classes where women - and men - can learn those arts which are designed specifically for the purpose of resisting attack. It is worth bearing in mind that people who have received training and know how to defend themselves are much more capable of remaining cool and handling an attack than those without any such background. And there is plenty of evidence to show that when a victim is able to handle the situation effectively, an attacker will turn and run. There have been many confrontations in which a lone woman has faced imminent danger of physical violence or rape but has escaped by keeping her wits about her, knowing how to deal with a man in an aggressive mood and how to try to talk him out of making the actual assault. In fact, some classes in self-defence measures put far more emphasis on this approach than on physical resistance.

Some simple guidelines you should always remember are:

1. On arrival home, first check the perimeter and do not enter alone if there is anything suspicious. Always have both home and car keys in your hand so that, if everything appears all right, you do

not stand fumbling for them in a pocket or bag, thus giving anyone who might have followed you a chance to come up behind you as you open the door.

2. Do not walk alone at night if it can be avoided.
3. Do not take short cuts through dark or deserted streets.
4. Walk near the kerbside so that you keep away from bushes, trees and buildings.
5. Always walk so that you face the oncoming traffic.
6. When walking at night, even if you are accompanied by a dog, always carry a torch.
7. Never hitch-hike.

If a confrontation does arise, how far can you go legally in defending yourself? It is a surprising question to ask but, as has been shown in a number of prosecutions, it is a fact that a victim of crime may be convicted and penalised for injuring the criminal in certain circumstances. In some countries, such as the United States, anti-personnel sprays are sold to women for the purpose of self-defence but such devices are quite illegal in the UK and only a few years ago a woman was prosecuted for importing one into this country. The legal position is a very difficult one and although the law, under Section 3 of the Criminal Law Act of 1977, permits the use of 'such force as is reasonably necessary in the prevention of crime', interpretation of the word 'reasonable' varies and a victim who is thought to have exceeded it may find that she – or he, of course – appears before a magistrate or a jury. In the street, or in any public place, for example, the carrying of daggers, knuckledusters, acid sprays, and sword sticks is illegal and most people can see the obvious reasons for this. Many women, however, think they ought to be allowed to carry disabling sprays in case of attack and would like to see the law altered. In fact, any such change could be far worse than the ban on these devices, since criminals carrying them for the purpose of attack could always plead that it was for self-defence.

How far one can go in one's own home to deal with a burglar is just as unclear, and court decisions in those cases where a householder has used a great deal of force against an intruder have not laid down any hard and fast rules about the legality of using blinding sprays or truncheons in the home.

These days there are, of course, different methods available to women for their personal protection which do not entail physical violence. A wide variety of 'screamers' are available for carrying in the hand, keeping in the handbag or leaving around in the home in handy places. As there are not yet any records of how many are in use or how many have been successful in preventing attack, their real value cannot be assessed. The devices, variously described as 'shriek' or 'shrill' alarms, some of which may be used in holsters with a lanyard to hang round the neck, are powered by replaceable containers of non-toxic, non-flammable gas. Most of them are activated by pressing the top down – Safe and Sound Products and Hoover are among the makers of this type – and at least one other, known as the 'Sound Grenade' available from Personal Alarms Ltd, is operated by means of a pull-cord priming system similar to a hand grenade.

Staff at some big stores, such as Marks and Spencer, have been issued with the devices and they are carried by many nurses, who have frequently been the target of rapists because they often have to walk in the dark from the hospital to a hostel. The device is pointed at the face of an attacker and emits an ear-splitting wail or shriek which serves a dual purpose. Firstly, it alerts the neighbourhood to the fact that someone is being attacked. Secondly, the volume of noise is quite intolerable to the human ear and causes the assailant to step or jump backwards involuntarily to escape it.

Specifically designed for the home, there are now the so-called 'panic' or 'personal' buttons which form a part of most modern burglar alarm systems or may be added as an extra in as many places as

Alarm system which, when considerable pressure is applied to the outside of the door, activates the alarm and thereby scares off the would-be intruder. The system also includes a 'panic' button (*above*) which gives protection against personal attack when the door is opened.

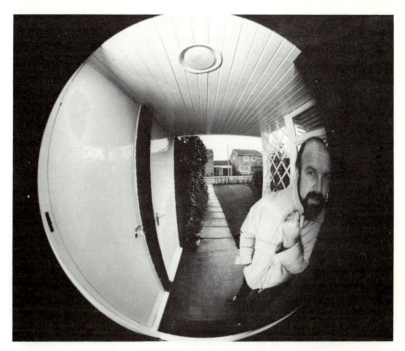

A doorviewer allows the occupant to see who has knocked on the door before opening it.

required. Situated at strategic points of vulnerability and risk such as inside the front and back doors, or in a bedroom or kitchen, they need only to be pressed to cause the burglar alarm to sound. Most of them are designed to operate even when the alarm system is switched off.

Opening the door to strangers is always a risky business. The use of a cheap and easily installed doorviewer allows the occupant to look out and see who is standing on the threshold before opening up – and the caller cannot see in. One range of such viewers – which cost only a few pounds, are easily installed by screwing together two parts after simply drilling a hole, and are available from most D-I-Y and hardware stores – is the 'Thuscan' made by Thunder Screw Anchors Ltd.

For those who would like something a little more sophisticated and offering a higher degree of protection, there are the door entry systems mentioned in an earlier chapter which are often found in blocks of flats and sheltered housing units. A number of companies, such as

Geemarc, Ademco Sontrix Ltd, AFA-Minerva and Status Electronics Ltd, supply a range of these products which vary greatly in both complexity and price. But they all have the virtue of offering good personal security.

The cheapest systems and the easiest to install are the audio ones which consist basically of a panel with one or more call buttons mounted on the wall outside the premises, linked by an electric cable to a wall-mounted or table-standing handset like a telephone inside the house. A built-in microphone and speaker allows the occupant to speak to a caller and to verify identification before opening the door. Some of these systems may be bought in D-I-Y kit form. Most of them have a facility to allow the introduction of an electric remote-control locking device for the door which is very convenient in situations where the occupied room is some distance from the door. Systems are available for homes with one or more entrance doors. Geemarc also supply an 'emergency' type for the elderly or handicapped, which enables the occupant to activate a siren and a red flashing light in the outdoor unit to indicate to neighbours or passers-by that all is not well.

Much more expensive is the range of video doorphone entry systems. These are a big step forward because they allow the occupier both to hear and to see the caller without that person even knowing that he or she is under observation. It provides the ability to check

Two-way door entry system showing the outdoor speaker panel (*right*) and wall-mounted telephone inside the premises (*left*). The front door is opened by remote control by means of an electric lock or release.

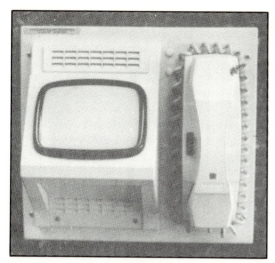

Audio-visual doorphone entry system which enables the occupier to see and speak to anyone standing outside the premises.

visually whether the caller is who or what they claim to be and enables the occupier, without indicating whether he or she is at home, to make no response if it is someone who is not welcome anyway. These systems, which incorporate a remote-control lock release, basically comprise a hidden camera with concealed lights for the outside, connected to an internal unit with a small monitoring screen giving both speech and vision. Some of the systems are so complex that they can handle installations with between one and 100 independent screen monitors.

Geemarc recently introduced a miniature but fully effective audio doorphone that can actually fit into the door itself. Called 'Telecall', it may be installed in either traditional timber or aluminium doors, is maintenance-free and may be either mains- or battery-operated.

When installed in multi-occupational residential premises, these access control systems, for that is what they really are, have been plagued with problems because vandals have still almost always had unrestricted freedom to go to an entrance unobserved and make nuisance calls, and this has naturally led to great resentment on the part of the tenants. On top of that, the security which the system was intended to provide has been negated by tenant activity such as wedging the main door open and jamming lock release buttons so that the doors remain permanently unlocked. After careful research into these problems, Status Electronics have developed a digital video access

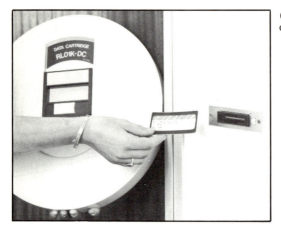

Chubb's programmable card access system.

control system with unique anti-vandal features and electronic circuitry which has been highly successful in providing a video doorphone entry installation which is able to withstand most forms of interference and as a result has much lower maintenance costs. It is also available in audio only form.

Panasonic Security Systems supply an installation which allows a connection of up to two entrance cameras and three room monitors which may be much more satisfactory than the smaller facilities offered by some of the systems if the customer lives in the larger type of house; and another for multi-occupation residential property allowing up to 841 room monitors.

Perhaps one of the most significant advances is the recent launch by Chubb Alarms Ltd of a programmable card access system of the type normally to be found only in industrial and commercial or other large complexes. The new system will, it is thought, have valuable application in some sections of the residential area, particularly the growing number of people who run a business from their home and have either an office or some other workplace there. Such activity often means that access needs to be controlled for the protection of valuable office equipment or plant. The conventional control systems have been costly and useful only for the supervision of large numbers of people, but the Chubb Entacard M300 can be programmed to accept anything between one and 300 cards at half the cost of any comparable system. If your business proves really successful, you can link ten card readers to give a capacity of 3,000 cards!

For those unfamiliar with access control systems, it should be explained that they work on the principle of a personally coded plastic card. This is either inserted into, or shown to, a 'reader' device at the entrance which has been programmed to identify the encoded data. If it is the correct identification the door is automatically opened, then closed and re-locked. The cards used for the new system are encoded with infra-red and believed to be the first of their kind in the world. Over 100 different codes are available, giving the system great security integrity. Its versatility is such that its applications cover the whole access spectrum from residential premises to power stations, from laboratories to private membership clubs. Anyone interested in this new development may be pleased to know, in view of the fact that much of the security equipment and systems used in the UK are imported from countries such as Japan and the United States, that this system is completely British and was developed and is manufactured by Time and Data Systems International based in Poole.

In the past, closed circuit television systems have been far too expensive and complex for domestic security use, but now low-cost miniature packages comprising a camera, a monitor screen, connecting cable and fixing brackets are available from a number of companies such as Ademco Sontrix Ltd. The versatility of such systems has meant that an increasing number of home-owners are installing this type of protection. One reason for the choice is that the camera picks

Video doorphone entry system showing the camera (*left*) which is installed outside the premises, and the monitoring screen (*right*) which enables the occupier to see who is outside.

up the image of an intruder before he actually reaches the premises and therefore it is impossible for him to cause any damage before he is detected. Secondly, because the camera transmits the image of the intruder to the monitor screen situated within the safety of the home, the occupier can not only watch what is going on but merely has to phone the police without the risk of going outside to investigate and facing the hazard of confronting an intruder – or intruders – who may be armed and violent.

Obviously, the main value of the system as far as the householder is concerned is during the hours of darkness, and therefore a source of light for the camera may be needed. Such systems when installed are linked with one or more floodlights for the surveillance of the rear of a house. This sudden illumination is in itself an important aspect, for no intruder is likely to try to continue a burglary in the glare of spotlights. Secondly, it must be remembered that there must be a device to activate the light source. This involves an exterior type of detector with a beam emission which, when broken by an intruder, causes the activation. The detector is linked to an alarm so that for the price of a low-cost cctv system the home-owner also gets the added protection of light and sound. Exterior detectors are plainly prone to a high rate of false alarms for the beam may be broken by anything from a swooping bat to a falling leaf, but this problem has been largely overcome by the introduction of units which transmit beams in parallel pairs. The pairs are so spaced that both must be broken before an alarm is triggered; small objects cannot do this but the intrusion of a human body does so.

These low-cost cctv systems are ideal for those people whose business and living accommodation are on the same premises. The camera may be positioned in the business part of the premises so that observation may be continued on that area by having the monitor in the living quarter. Ademco Sontrix supply a special connector which enables the camera to be linked to an ordinary television set if that is required for any special reason. For night-time observation on the monitor if an alarm is triggered, the camera could, for example, be linked to a portable television in the bedroom.

There is another form of intrusion which, as many women know from experience, is not physical attack but is just as frightening in many ways and by its very nature may be very distressing to the victim: the obscene or nuisance telephone call. It is disturbing to note that the number of such calls is increasing – according to the Home Office publication *Crime Prevention News* there were over 150,000 complaints in 1980 and the incidence of such calls is increasing at the rate of ten per cent every year.

It is difficult to deal with this sort of intrusion because of the problem of tracing the calls. But British Telecom has been doing a great deal of research on developing improved techniques which, it is hoped, will lead both to identification and successful prosecution of the culprits. In 1980, two test trials were run in Glasgow and Nottingham to try to assess the best methods of dealing with them. During the trials, over 12,000 such calls were received and nine out of ten of them were satisfactorily concluded by giving advice to subscribers on the best way to deal with them.

As a result of the trials, British Telecom prepared a leaflet which is available from their general managers' offices throughout the country. The advice it gives to women includes the following:

1. If you receive an obscene or nuisance call, hang up gently without showing emotion, and in most cases the caller will not ring back.
2. Make sure that children or babysitters answer the phone correctly, without giving away private information such as 'Daddy won't be back until . . .'.
3. Women whose numbers are listed in the telephone directory are advised to use only their initials, and not Miss or Mrs.

The British Telecom leaflet has a tear-off slip so that a subscriber can record details of obscene or nuisance calls in case she wants an attempt to be made to trace the caller. But it must be made clear that British Telecom can consider trying to do this only if the customer gives an undertaking to support a prosecution by giving evidence in court if the trace attempt is successful. Many women find it too distressing to have to describe their experience to a court and to undergo cross-examination by the defence.

If nuisance or obscene calls are made to a subscriber persistently it is possible in some areas for the operator to intercept all incoming calls. In some cases a change of telephone number, if there is a spare one available, combined with a decision to go 'ex-directory' may be the only answer.

8 Car security

At least 300,000 cars are stolen in the UK every year. Some of them are taken by so-called 'joy riders' and are later recovered, often damaged by accident or by being vandalised, but many disappear completely without trace. The most frequent types of offence recorded in England and Wales in 1982 – the last year for which figures were available at the time of writing – were those of theft from a vehicle, totalling 449,000, and those of theft and unauthorised taking of vehicles, which totalled 351,000. In fact, these two crimes account for about a quarter of all recorded offences. These figures are hardly surprising, however, when you consider that a kerbside check carried out in six major cities revealed that an average of one car in every four was insecure. And yet a few simple precautions may well cause the would-be thief to try to find an easier target.

For many people, their car is really an extension of their home. They use it for business, for pleasure, for taking the kids to school, for shopping expeditions. It would seem natural, therefore, to ensure its security and the protection of its contents in the same way as the home, and to maintain that security at all times, whether the car is parked in the street or left in your garage.

Just as with basic security for the home itself, there are certain basic rules to keep the car safe:
1. Always remove the ignition key when the vehicle is not being used, even when it is garaged.
2. Do not keep your log book, driving licence or insurance documents in the car.
3. Most cars have steering column locks; if yours does not, see that it is immobilised in some other way.
4. Whenever you leave the vehicle, even for a few minutes, lock the doors and fasten all windows, including the quarter lights.

5. Never leave the car unattended with luggage on the roof rack.

6. Don't leave the car with valuables in view – remove them or lock them in the boot.

7. If you do have valuables in the boot, remember to lock it.

8. When parking at night, always choose a well-lit place.

Steering locks were considered to render a car virtually theft-proof when they were first introduced, and for a time that was the case. But they are no longer as safe as they were because thieves have now found how to overcome them. A survey carried out in 1981 by the police forces in Merseyside, Kent and Lincolnshire to study the methods used by thieves to steal cars fitted with steering column locks showed that in 89 per cent of all such cases the lock was overcome either by the use of a key or by 'jiggling' the mechanism. Force was rarely used.

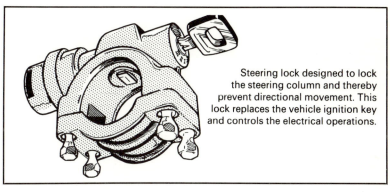

Steering lock designed to lock the steering column and thereby prevent directional movement. This lock replaces the vehicle ignition key and controls the electrical operations.

But more sophisticated methods to break the locks were used mainly by professional criminals and accounted for only one in ten of all cars fitted with steering column locks which were stolen. If you have a steering column lock on your car it is advisable to have some other form of protection as well, especially if it is of an older type and is at all worn.

If you ride a motorcycle or a moped, remember that they are just as likely to be targets of the thief as the car so never leave them, even outside your home or in the drive, without first ensuring that they are immobilised or securely locked to something immovable with a chain and padlock.

Many people, particularly when out shopping, do not realise that the modern thief has modern targets in mind. He may not bother about anything bulky or conspicuous but will usually look instead for something much more valuable to him and very easily saleable if he does not wish to use it personally. That target is a cheque book, cheque guarantee card or a credit card, and if he finds them all together – many people leave them in the glove compartment – then he really is in luck. One of the first things the car contents thief does is to break open that glove compartment.

How many people realise that valuables in a car include radios, hi-fis and especially CB sets? If they are not fixtures, take them out and lock them in the boot when leaving the car. If that is not possible, locking kits are available to make them more secure.

Have a locking petrol cap fitted because, with petrol at its present price, it is often siphoned from the tanks of parked cars.

Over and over again drivers have returned to their parked vehicle and found that wheels and tyres are missing – even when the vehicle is parked in the drive at home. This sort of theft may be prevented by fitting locking wheel nuts which are cheap and easy to fit.

The principle upon which another ingenious device is based is a simple one: thieves will not touch a car that cannot be moved. So the Wheelok, manufactured by Lionweld, clamps the wheel and stops it turning – in the same way as the clamps which are used by the Metropolitan Police to deter illegal parking. As the clamp prevents the wheel nuts from being reached, owners who can afford the extra expense can use one on each wheel to safeguard their wheels and tyres.

If you own a new car it will be a good investment to take special precautions against the professional thief who may well belong to a car-stealing gang with a sophisticated organisation waiting to change the registration number, then remove all identity marks and sell the car, probably abroad. Such precautions include having the registration number etched onto all the windows of the vehicle. A thief would then have to buy a complete set of the glass windows and, apart from the large expense involved, it would add considerably to his risk and would alert the supplier. There are a number of companies supplying these etching kits, including the Etching Transfer Co with their 'Automark' product.

If your car is more than ten years old you may think that you don't have to bother about things like this – but you could not be more wrong. The risk of an older car being driven away by 'joy riders' is several times greater than if it was a new model. Old cars are easier to steal, harder to protect and less conspicuous. So make certain all the locks work and are not badly worn. And make certain you secure the quarter lights whenever you leave the vehicle because that is one of the commonest ways of entering an old car.

Cars, like houses, of course, cannot be totally protected against a determined thief. But in addition to the basic precautions described above, there is a wide range of alarm systems, immobilising devices and special security locks which can be fitted as extras and which offer a high degree of security. Alarm systems of varying degrees of complexity are offered by firms such as Simba, Waso, Motolarm, Gamma Electronics, Selmar and Davco Instrumentation. The Simba Inerti-alarm provides comprehensive alarm and immobilisation for any make of vehicle. Sensors placed within the body detect any attempt to force locked doors or windows, or to make entry through the boot or the bonnet. This system is manually operated by means of an external keyswitch and may be linked to the more usual method of protection by using switches in the doors and other openings. Closed circuit wiring is used to prevent interference by a thief.

The 'Black Box' system introduced by Waso is aimed at the lazy or forgetful motorist and is automatically set as soon as the ignition switch is turned off. The ignition is dead until the unit is deactivated and any form of tampering causes the alarm to sound.

Kidnapping of executives has led a number of car alarm firms to introduce special features to guard VIPs while travelling in their vehicles. Davco Instrumentation manufacture one such system which, by use of a special hidden switch near the driver's seat, causes the vehicle to splutter to a halt after a few moments, giving the impression that it has broken down. It is claimed that this simulation causes immobilisation without creating any suspicion. The car cannot be restarted without the use of a second switch which is hidden in another part of the vehicle, such as the boot. Davco say that this system could be of great consolation to the lone woman driver faced with forcible intrusion into her car by an attacker, who would be unlikly to persist in his intentions if he were left stranded with his victim in the middle of the road.

Since most of the alarm systems which rely on noise to raise the alert work off the car's horn, it is essential to ensure that there is an automatic cut-out and re-set function. Of course, it is not necessary to have an installed system as there are a number of D-I-Y kits available, some of which are quite easy to fit. But, as with household alarms, the degree of protection offered varies considerably and the more sophisticated and comprehensive the system is, the more it costs and the more likely it is that you will require professional installation. It obviously depends on how much you value your car and how much of a blow its loss would be.

For those who do not consider an alarm system a good investment anyway, and for those who cannot afford one but to whom use of their car is essential, there are effective and cheap devices available which, like the familiar 'Krooklok', hook over the steering wheel or are attached to handbrake or gear lever to physically deter any attempt to drive it away. A professional car thief, determined to take your vehicle, may easily overcome such obstacles but the 'joy rider' or opportunist thief is not going to bother to waste time on trying to circumvent them.

It is still a fact, which few people either remember or bother to put into practice, that the simplest way to prevent your car being stolen, and one which costs nothing, is to remove the rotor arm from the distributor and take it with you when leaving the vehicle. But, of course, this and other owner-interference methods are messy, depend on some knowledge of car mechanics which many do not possess, and are not always practical or convenient.

Motorists who believe that the professional thief knows just how to overcome conventional safeguards might consider the novel idea of a firm called Security Equipment Installations. They supply locks with removable barrels for car doors and boots so for those thieves who rely on lock picking there is nothing to pick. Or you might consider a system which will let you know if someone is tampering with your car when you are somewhere else, even up to eight miles away. A company called Car Sounds supplies a kit consisting of a transmitter which fits into your car and a receiver or 'pager' which you carry with you. If anyone interferes with the car while it is parked you get a 'bleeper' warning reinforced by a light.

Modern radio communications give drivers protective cover by enabling them to call for help if they are attacked or are anxious about suspicious circumstances, such as a persistently following car when they are carrying valuables. This is the two-way relay telephone service linking the driver by car phone to one of a nationwide chain of receiving stations. The three main companies in this field are Aircall, Securicor-Granley and Network.

The number of people who own caravans is increasing and, whether parked in the front garden, on the drive or on a remote site, caravans and their contents are now frequently being stolen, vandalised or used by squatters. Always remember, therefore, to extend your security measures to include your caravan. All the alarm companies supply units which offer a high degree of protection for both trailer and static types of caravan. Burglar alarms similar to those used in the household itself are available and may be battery-operated or connected to a mains supply, installed or free-standing.

It is a good idea to fit extra locks to your caravan and if necessary to replace the maker's locks with better quality ones. Do not leave valuable equipment where it can be seen through the windows; and draw the curtains, anyway. Engrave the registration number or your postcode on all the windows. If your caravan is of the trailer type and a key-operated hitch lock is not included as standard, have one fitted to prevent the caravan being towed away. Remember, however, that trailer couplings are not universal so ensure that you get one that is compatible with yours – and if it is of the type which is dependent on a padlock, make sure that this is sturdy enough to resist hammer blows.

Another device which prevents illegal towing away is the leg lock. This secures the legs in the downward postition and is an easy D-I-Y job. Wheel locks not only prevent illegal towing but protect the wheels and tyres from removal as well. Similar precautions are necessary for a boat on a trailer. Outboard motors are a particular target for thieves so make sure they are bolted, chained or locked in position – or remove them altogether when the boat is not in use.

Obviously, all these precautions are especially important if the caravan is your home all the time, whether it is on a permanent site or whether it is a mobile home sited in a different place from time to time. When it is left unoccupied it is particularly vulnerable, because thieves know that it is a home containing valuable items just as if it was a brick-built structure. Alarms which operate loud sounders are very effective as safeguards because the possibility that neighbouring van dwellers have been alerted will deter most thieves.

Companies which produce smaller types of alarms for the caravan and the boat include Peak Technologies, who supply a completely self-contained portable system with rechargeable batteries. Trends Security Alarm Systems also produce individual battery-operated door and window alarms which need no wiring or electric mains connections, and are available from many stores, D-I-Y outlets and hardware shops. There is even a safe which is specifically designed for the mobile home, made by a firm called Stoughton Lodge (Security) Ltd.

Finally, there is one important aspect of home security relating specifically to the family who go away on holiday and at weekends in their caravan which is normally kept parked in the front garden or drive. Few people realise that the fact it is missing from its normal place indicates to an observant burglar that the family must be away and the house left empty - you have been warned! The only security step you can take to deal with this situation is to make sure that a neighbour keeps a careful eye on your property.

The theft of bicycles is reaching epidemic proportions, and thousands of them are stolen all the time. Never leave your machine without locking it: use a heavy-duty chain and padlock or a wire hawser type of lock and pass the chain right round the frame and some other immovable object. Tell your children of the danger and make sure they never leave their cycles lying around. Keep a record of the serial numbers and descriptions - particularly any distinguishing marks - of your own and your family's cycles. And, once again, engrave your postcode and house number on each machine under the main frame at the bottom.

9 Sheds and garages

Most people keep a grand array of tools in their sheds for use in the garden and for all those jobs that need to be done around the house. Then, of course, there are the ladders – some of them long enough to reach the bedroom windows. Few home-owners recognise the risks involved in presenting a burglar with such a tempting range of housebreaking tools – for that is what they are in the hands of the villain, quite apart from the fact that he will steal them as well.

Keep all sheds and garages securely locked up. Such buildings are obviously vulnerable anyway, but that is no reason to make it easy for the intruder. If possible, keep any ladders in the garage since it can usually be made very secure, but even in there lock them to a hook or support if you can. If you have to keep them in a shed – which is probably at the bottom of the garden and hidden from view – it is essential that you lock them to a bench or hook. If you are forced to leave them in the garden, lock them to a stout post or something similar.

Just as in the choice of locks for the house, remember that any old padlock just will not do. Although they may cost a bit more, high quality padlocks of various kinds are available, some with five- and six-lever operation giving a big differs permutation. Some of these locks are what are known as 'close shackle', which means that the closing part of the padlock is protected by a hardened steel shutter designed to resist cutting or sawing. The padbars which are the fittings on the door should, ideally, have no exposed screws or bolt heads on the surface.

The security of the garage is most important. Many home robberies have been made easy by access through an insecure garage door, especially if the building is attached to the side of the house. Once inside the garage, a burglar can get to work undisturbed because he is out of sight and often any noise he makes will not be heard.

Many people think that once their car is in the garage it is safe. They are wrong. There are hundreds of recorded thefts of cars from household garages, and of the car's contents even if the vehicle itself is left.

Many garages contain much valuable property in addition to a car – for instance, the freezer with its load of costly food is often installed there. If there is an up-and-over type door with a special built-in locking system there will be good security anyway, but if it is held by means of a single bottom exterior fitting then extra locks should be added. Of course, many burglar alarm systems include an alarm for the garage as well. Even if they do not, any good installer will give advice and make the necessary addition. But in any case the conventional type of double garage doors should have a surface-mounted rim automatic deadlock, and each door should have a surface-mounted bolt at the top and the bottom.

Take special security measures if you have a front porch which is of the enclosed type, and especially if it has opaque glass or is of any form of construction which would give cover to an intruder who could shut himself inside and work at leisure on opening the front door into the house. There are many people who take steps to ensure that the front or final exit door is properly secured – or so they think – and then leave the porch door open or insecure. But in this situation the porch door, of course, has become the final exit door and, if it is fully glazed as the majority of porch doors are, it has become very vulnerable.

Ideally, glazed porches should have clear glass so that the actual front door is visible, and the locks should be of the same high-security standard as those on the front door itself. The glass near the lock may be protected by using an interior fitted decorative metal grille. Or, better still, the whole glazed area may be given the security of an internal application of shatterproof film or a secondary interior pane made from a virtually unbreakable glass substitute material of which there are many proprietary brands on the market.

Exterior security applying to sheds, porches and garages is particularly important when you are away from home on holiday – families often leave without first checking that everything is secure. Open or unlocked windows have almost invited thieves into the home, while insecure sheds and carelessly left ladders have given them the means of breaking open what is locked up and provided access to low roofs for entry to upstairs windows.

If it is at all possible, do not make a big display outside the house when you are going away: it may be possible, for instance, to pack the luggage into the boot of the car while it is still in the garage rather than do it in the street. Burglars are always on the look-out for the tell-tale signs which indicate that you are going away – or that you are away. Do you remember to cancel *all* deliveries before you go on holiday? Milk bottles and newspapers delivered and then left to build up day after day tell the burglar just what he wants to know. Do you ask a trusted neighbour to keep an eye on your house or flat while you are away? If anything seems wrong or anything suspicious is seen or heard, the neighbour can investigate or, when necessary, call the police. Do you leave a key with someone you can trust? It may be a neighbour or a relative living nearby, for if the police are called to your home they must have a means of entry. Unsolicited leaflets and uncollected mail sticking out of the letterbox are a complete give-away – make sure that a neighbour takes them out, or pushes them right in if it is an enclosed letterbox so the contents are not visible from the outside.

Remember that by using one of the special timeswitches such as Superswitch while you are away you can programme it so the lights are automatically switched on and off at different times to give the impression that there is someone at home. Do not use the same light in the same room all the time.

Do you leave an address or telephone number with a relative or neighbour – even if you are touring it may be possible to give a forwarding address – so that you can be contacted in a grave emergency such as fire? Do you tell your local police that you will be away from home? A patrolling officer will keep an eye on your home any time he is passing and may notice anyone acting suspiciously before any harm is done.

Finally, for safety reasons, make sure that all electrical items such as television sets are switched off and unplugged.

10 Danger from fire

In every week of the year, there are more than 1,000 fires in dwellings in the UK and about 14 people die each week as a result. About another 100 are injured. That means a death toll of at least 700 people each year, and there are some 500,000 minor or 'near miss' domestic fires causing injury to another 6,000 or so. Many of those injured are children who are badly burned, spend months in agony in hospital undergoing surgery and are scarred or maimed for life. Fire presents a danger to every home but few people take any precautions against it. Although only one in six of the total number of fires in the UK takes place in a dwelling, three out of four of the total number of deaths and two out of three of the injuries are caused by them. It is a sobering thought that many of these tragedies could be prevented by ensuring that a few simple measures are taken.

Over one-third of all fires in dwellings are the result of a cooking accident, often involving a chip pan catching fire. But these kitchen stove blazes need not be lethal: it is because people do not know what to do when it happens that a dangerous situation develops. The first thing is to keep calm. The second is simply to turn off the electricity or gas and thus remove the heat source. Then you need to smother the flames to starve the blaze of oxygen so it will go out. A fire blanket – an inexpensive cover used to smother the flames – should always be kept in the kitchen close to the cooker. Remember, do *not* throw water on the flames as this will only make the blaze worse.

Smokers' materials, including lighted cigarettes and pipes, are the second most frequent cause of fires in the home and are the worst cause of fatal conflagrations. It is believed that the reason for this is that a large number of these fires start when people go to bed, for a smouldering cigarette end may go on burning down the side of an armchair for hours before it flares up – so the fire grows while the family is asleep upstairs.

There are, of course, other causes of fire such as an electrical fault developing under floorboards or in the loft. A slow rise in temperature during the night may not be detected and during the day a similar situation may not give rise to suspicion because increasing heat might be attributed to another cause such as cooking, or the fire might be in an unoccupied part of the premises anyway. A build-up of smoke or fumes may also be undetected for similar reasons and may be allowed to continue until it suddenly culminates in a rapid and disastrous outbreak of fire.

One way to provide early detection is to install one or more smoke detectors in the home. They are self-contained, easily installed and will give you adequate warning. Smoke is usually the killer and renders people unconscious long before there is any question of them being burned. Most of the materials used in modern furnishings produce more smoke than flame. Smoke detectors, which may be ceiling- or wall-mounted, react very early to smoke production and emit an earpiercing sound sufficient to wake even the heaviest sleeper. Depending upon the size and layout of the premises, more than one detector may be necessary to give comprehensive protection but very often a centrally located single detector may be quite sufficient. All fire brigades give a free fire prevention advice service so you could easily find out just what you should do by telephoning your local fire brigade headquarters or the divisional headquarters. Most insurance companies will also give advice to their customers.

In addition to a fire blanket and a smoke detector, a small fire extinguisher should also be purchased and kept in a handy place for dealing with fires which may occur from other causes and which are quickly visible and detected. Fire extinguishers contain different kinds of extinguishing agents for use with different types of fire, so get some advice when buying a unit otherwise you might make the situation worse. Fire blankets, smoke detectors and fire extinguishers may be bought separately or as a combined kit, with the instructions enclosed.

Just as dangerous as the slow build-up of fire is the build-up of gas from a leak in the system which may go quite undetected – especially if premises are unoccupied – until there is a violent and catastrophic explosion. In the past few years there have been a number of serious explosions resulting in the loss of several lives and people have seen their homes completely wrecked. Banham Locks and Alarms Ltd have just introduced a revolutionary device which will put an end to

Ceiling-mounted smoke detector.

this tragic situation. It is fitted over the exterior main inlet gas pipe and if there is a leak it detects it instantly and automatically turns off the supply.

The construction and layout of many homes may make escape difficult or impossible in the event of a fire breaking out, especially if it starts downstairs and quickly engulfs the staircase. There are means of escape available such as fold-away ladders which can be used as a fire escape at upstairs windows at the rear of the house.

Burglar alarm systems frequently include either one or more smoke alarms or the option of including them as part of the system, and this may be particularly valuable if your alarm is connected to a central station where an operator will receive the alarm signal and call the fire brigade.

For more comprehensive fire coverage, heat detectors are also available which will sound an alarm if the temperature in the premises rises above a permitted maximum. They are probably most useful in the kitchen but are not very often necessary in the private home.

Gent Ltd, who claim to be the biggest manufacturers of fire alarm equipment in the UK, have introduced a low-cost fire alarm kit including both smoke and heat detectors, which has been designed for use in small, multi-occupational residential premises, which provide the organisations responsible for such premises with the chance to install comprehensive fire protection. This system is of great value to welfare and religious bodies with limited funds.

General advice issued by the Home Office and the Scottish Home and Health Department stresses that everyone in the home should know what to do if fire does break out. They give the following instructions:

1. Close the door of the room where the fire is to help prevent the spread of poisonous fumes and also to restrict the fire.
2. Alert the household and get everyone out by the safest route. If you live in a flat do not use the lift because the wiring could burn out and trap you between floors, or you might unwittingly descend into a

Fire alarm kit designed for small residential premises.

raging furnace since you are unable to see out until the doors have opened.

3. Alert the neighbours and call the fire brigade by dialling 999. You do not have to put money in a public call box. Do remember to give the full address.

4. Try to reduce draughts that may fan the fire – close all doors and windows, even in rooms away from the fire, if this can be done safely.

To avoid an outbreak of fire the advice is as follows:

1. Make sure that chip pans are not more than half full and *never* leave them without turning off the heat. If you can afford it, replace your old open pan with an automatic fryer.

2. Make sure that portable or free-standing heaters are not placed close to furniture or furnishings and take care that nothing is left where it can fall on them.

3. At bedtime, switch off television sets, radios and portable electric heaters and pull out the plugs. Check ashtrays for burning cigarette ends or pipes.

4. Close all internal doors – this may confine the fire to one room and will certainly slow its spread.

There are two important don'ts concerning fire hazards: do not smoke in bed as it is a major cause of fire in the home; and do not leave children alone where heating and cooking appliances are in use or leave matches where children can reach them.

Finally, keep an eye on the elderly and make sure they take sensible fire precautions.

11 Domestic monitoring and lower premiums

In the spring of 1983 the Birmingham-based company of ADT Security Systems entered the domestic security market in a big way with a system which takes care of monitoring refrigerator and freezer mechanisms as well as incorporating a complex array of security and fire protection devices. Security means peace of mind as well as material protection of people and property, which is why ADT have included in their 'Safewatch' domestic security system features well beyond the public expectation of a 'burglar alarm'.

There is nothing new in this concept, nor is this the first company by any means to incorporate this type of auxiliary alarm service in their system, but it is a comparatively recent innovation because monitoring has traditionally been geared to the management of industrial and commercial premises. But it demonstrates how security is going to develop in the future and what the consumer can expect to be offered. Most of the leading alarm companies, and many of the smaller outfits as well, now provide similar monitoring facilities for the home.

When Intercept Alarms introduced their alarm system, one of the special auxiliary benefits included the monitoring of the temperature of a tropical fish aquarium where a failure in the absence of the owner might mean the loss of a collection of very valuable fish. And with the creation of companies such as Britannia, Pyke International, Central Monitoring Services, and Keyco Custodian, the entire UK is now covered by a series of monitoring services based on specially designed computerised central stations.

The latest addition, claimed to provide the most advanced central station computer monitoring service in the western world, and backed by a group of top companies including the General Accident Fire and Life Assurance Corporation, is Monitron Computer Monitoring Services Ltd, a London-based company. It offers monitoring for every conceivable situation, is geared in particular to the domestic field and, except in the case of fire and personal attack alarms where there must be no delay in response, endeavours to check the validity of all calls before the appropriate service is alerted.

A central station is a control room which acts as a terminal for alarm signals from the protected premises, and is manned day and night by trained operators. A communicator located on your premises, either attached to or connected to the telephone or, in the case of the new style of phones now being produced, incorporated as part of the machine itself, transmits a signal which is automatically picked up on a receiver at the central station. All relevant information, such as the nature of the problem, the address and the telephone numbers of people such as key holders who may be contacted, has been pre-recorded and is shown on a visual display screen. After carrying out certain instructions intended to establish whether or not the call is a genuine alert, the operator notifies the appropriate service, such as the police or the fire brigade.

Items such as freezers, fridges, and fish tanks are fitted with buzzers which sound if anything goes wrong – it does not matter whether the equipment is in the house or the garage – and the signal is transmitted in the same way. Pyke International include in their range of auxiliary alarms a float switch which will sound if the basement of the protected premises is flooding, which can be a very useful safeguard in certain situations. For dealing with these auxiliary alarms, where the services of police or fire brigade are not needed, special arrangements are made with key holders.

It has been mentioned already that some alarm systems, such as those designed for the care of the elderly and the handicapped, are either monitored continuously, or automatically at regular intervals irrespective of an alarm condition, by means of signals that are electronically analysed to establish that the system is working properly.

For years now, insurance companies have been heavily criticised for not offering premium discounts to those who introduce intruder and fire alarm systems into their homes, instead of merely increasing premiums to meet ever-increasing losses while exhorting their customers to take adequate precautions. At last there are signs that inducements in the way of lower premiums to encourage home-owners to take such security measures are becoming commonplace. Insurance companies who fail to do this may find in the future that they will lose business if they do not fall into line.

The involvement of General Accident with the operating central station of Monitron Computer Monitoring Services – the first time there has been such an association in the UK – marks a significant change in outlook. Monitron are offering a 20 per cent premium discount on insurance to any domestic alarm system user linked to their service. A group of insurance companies, all of whom are members of the British Insurance Association, are behind the offer.

Until the spring of 1983, apart from local package deals arranged through insurance brokers in one or two areas of the country, it was still general insurance policy to insist that, in an ordinary household policy, the proportion of the premium devoted to burglary loss was so small that no premium discount large enough to be attractive to the customer could be made. Then Hoover broke the barrier by announcing that, in conjunction with Life and General Ltd (Insurance and Financial Services) of Marylebone Road, London, they had reached agreement with the Economic Insurance Co to provide recognition for the householder who installed Hoover Firecheck and Hoover Thiefcheck appliances. A premium discount of 12 per cent on buildings and contents cover was offered when either of the two Hoover systems was installed, and this discount increased to 15 per cent if both were installed. Furthermore, a five per cent discount was offered if Hoover smoke detectors were installed, and special benefits under the Hoover Privilege Home Insurance plan were available.

Since then, a large number of special premium discount offers have been made to householders installing alarm systems. Vigilante, the domestic security division of the Reliance Group, linked up with the Bradford-Pennine Insurance Company – part of the Phoenix Assurance Group – to provide reduced premiums on home contents insurance for those spending more than £30 on the security company's hardware.

The Cornhill Insurance Group give premium reductions to householders installing Chubb alarm systems: 15 per cent on contents and personal possessions insurance for alarm protection up to British Standard 4737; $17\frac{1}{2}$ per cent for alarm protection up to BS 4737 plus smoke detectors, extinguisher and fire blanket; five per cent on a range of door and window security locks plus fire blanket or extinguisher or smoke detector; and 20 per cent on a combination of these products. On the buildings insurance there is a $16\frac{2}{3}$ per cent discount for installation of smoke detectors, extinguisher and fire blanket.

The country's largest household insurers, Sun Alliance, give a ten per cent premium reduction but only where there is minimum cover of £20,000. However, they have no links with any alarm company and do not insist on the installation of an alarm system at all – 'good overall security' will be sufficient to qualify and their surveyor will offer practical advice.

Although the principle of reduced premiums for security-minded customers is good and there would seem to be no really valid reason why this sort of discount should not be the general rule, perhaps the British Insurance Association's claim that the amount is too small in cash terms to have any real effect is also right. For in financial terms the saving of a few pounds on an insurance policy is not really likely to tempt householders to spend large sums on the purchase and instalment of expensive security equipment.

The most interesting insurance cover development – claimed to be the first of its kind – is Home Burglary Distress Insurance. It has just been introduced by the Bristol-based brokers, K.T. Jarrett Ltd. The heartache and trauma caused to many home-owners as a result of burglary often culminates in the need for special medical treatment to assist in mental rehabilitation, or produces a compelling desire to move home altogether once the security and privacy has been violated. The new policy covers removal expenses and medical treatment required as a result of a burglary. Already, Notecalm Alarms Ltd of Bedford have made an arrangement with K.T. Jarrett for special terms to be offered to home-owners whose property is guarded by a Notecalm system. The distress insurance policy, costing £2 per £100 insured, is available separately or as part of the Jarrett household protection scheme arranged with Notecalm.

The British Insurance Association have revealed that a quarter of all householders have no insurance at all for the contents of their home, and that many more insure for amounts that are too low. It is disturbing to think that some people will be unable to replace possessions damaged or lost through fire, flood or theft when they could have been protected against these events and many others by arranging adequate home contents insurance.

Whether you own or rent your home, you need insurance cover for the contents. Such policies normally include furniture, furnishings, household goods, kitchen equipment and other appliances, food and drink, televisions and radios, clothing, personal effects and valuables such as jewellery and personal money up to certain limits – although boats, caravans and motor vehicles are insured separately. There are two types of cover available: 'indemnity' and 'replacement as new'. With the first type, you will be paid the cost of repair or replacement less an amount for wear, tear and depreciation. With the second type, you will be paid the full cost of repairs and replacement. The property you can insure on this basis varies from company to company.

Every home-owner and tenant should go into every room in his home and into the garage and shed and make an inventory of every item, estimating what each one would cost to replace at present prices. He should then add up the costs in order to calculate the sum necessary for which to insure; remembering to keep the inventory in case it should be needed. For items covered on an indemnity basis you must deduct an amount for wear, tear and depreciation to arrive at a correct total. Where it is difficult to establish the figure yourself – in the case of valuables or antiques, for example – a valuation by an expert may be necessary. And remember that there is usually an upper limit on the value of any one article such as a work of art, an ornament or a piece of jewellery, and sometimes items such as video machines. Insurers may agree to raise those limits if you give them full details.

Because values are constantly changing, it is necessary to keep the sum insured up to date. Many insurers help by linking your insured sum to the Government's Retail Price Index. This means that your insured sum is automatically changed every month in line with any movement in the Index, and there is no charge for an increase between renewal dates. You should always keep bills and valuations because they will be very helpful if you ever have to make a claim.

You must also have adequate cover for your home buildings. This covers not only the structure of the home itself but permanent fixtures and fittings too and, because of the danger of damage or complete destruction by such things as fire and explosion, this must be adequate.

Free leaflets giving detailed advice and check lists on both contents and buildings insurance may be obtained by sending a stamped addressed envelope to the British Insurance Association, Aldermary House, Queen Street, London, EC4N 1TU.

12 What to do if you are burgled

The first thing to remember if you find that your home has been burgled is to try not to panic, no matter how upset you may be by the scene of devastation facing you. It is often the woman of the home who makes the discovery, probably on her return from shopping. So what do you do as you step inside the front door and realise that something is seriously wrong? The hall itself, depending upon the layout of your home, may be all right, of course, and it may be only when you reach the living room that you are faced with the reality. It looks as if an earthquake has hit it. Suddenly you feel very sick, and you are frightened. You are almost certainly alone.

For all you know, the intruder or intruders may still be on the premises. Turn quickly and quietly, and leave your home, closing the front door behind you. Go at once to your next door neighbour's home or to the nearest neighbour who is at home. The first thing to do there is to telephone the police and tell them what has happened. If you have remained in your own home and rung from there, do not on any account touch anything.

It is almost instinctive to go straight to where you keep your most cherished possessions, perhaps your jewellery, or some spare cash, to see if it has gone. Don't! Resist the temptation. It is an automatic reaction to straighten perhaps a chair that is on its side, or to pick up something lying on the floor and replace it. Don't! There may be fingerprints or palm prints or some other vital clue which will help the police to identify the intruder, and if you touch anything you will destroy them. The police will tell you what to do and that will be to leave everything just as it is until an officer arrives at your home.

Once the fingerprint officer has finished you will be asked to look around and check to establish what has been stolen and to make a list of those items. You will need one for the police, one for your insurance company in order to make a claim, and one to keep for yourself. Beside each item you should put what you consider its value to be.

You must notify your insurance company at once and you will be asked to send in a claim form with details of the part of the premises from which each item was taken and to give details of how entry was obtained. If the intruder has broken through a window or door which is now unsafe as a result of damage, you are entitled to make immediate arrangements to have the necessary repairs carried out so that the premises are secure and then ask for reimbursement later as part of your claim.

Most residential burglary losses are not large enough to warrant a check being carried out by the insurance company, even though they may dispute the amount of your claim. But if the loss is comparatively high, or involves the theft of an antique or some other item whose real value may be difficult to determine without a valuation, then the insurance company may insist that your home is visited by a burglary surveyor or loss assessor who will ask for more details before any settlement is arrived at. You may be asked to take steps to improve the security of your premises. If there are any weaknesses they will be pointed out.

So serious is the emotional effect of burglary on many victims that a whole chain of what are known as victim support schemes have been formed throughout the country. They work in close co-operation with the police and do everything possible to reassure the victims and help them through the initial trauma. If you have no one to whom you can turn, the scheme can put you in touch with a wide range of agencies which can help, such as the social and medical services.

The National Association of Victim Support Schemes, whose head-quarters are in London (the telephone number is 01-582 6500), was started in 1980 with funds contributed by the Home Office, the De-partment of Health and Social Security and voluntary services. Reg-istered as a charity, its annual report published at the end of 1983 disclosed that in the previous year more than 40,000 victims of crime were visited and provided with help and advice. Helen Reeves, the national officer of the Association, has pointed out that, despite a popular belief that its work is of value only to the old and vulnerable, the British Crime Survey clearly showed that although the elderly and infirm feared crime the most, they were the least likely to become victims. Young couples setting up home for the first time might suffer just as much from a burglary as elderly people, especially if they have not arranged their first insurance.

There are at least 130 schemes now in operation, covering more than 150 districts. Many of them receive some form of Government grant and a number employ paid co-ordinators whose job is to liaise closely with the police each day. Whenever it is felt necessary a volunteer visits particular victims. If you are a victim and need help and advice, the Association will do all it can including giving assistance in claiming injuries compensation if that is required.

Summary: The rules you must remember

It is no good fitting adequate locks to your windows and doors unless you use them. So remember to see that all access points are properly locked every time you go out.

It is a waste of time having a burglar alarm system installed if you do not switch it on. Remember to see that it is in operation whenever you go out shopping and whenever you go away on holiday.

Fitting a safety chain or limiter on your front door is useless unless you keep it on every time you open the door until you have identified the caller. Remember not to release it at all if the stranger is someone you do not wish to admit even if you have made an identification.

Try to cultivate that sense of awareness, of security-mindedness, already described. Remember that one of the nasty facts about everyday life today is the bogus caller. Especially if you are elderly, or handicapped, or a woman alone, do not admit anyone unless you know them or are quite satisfied they are what they purport to be. The man or woman who claims to have come to read your meter may be someone you recognise as a result of regular visits. But if it is a stranger, make the person produce an identification document to prove he or she is genuine. And remember, do not give the document just a cursory glance: make sure it is an official document.

If the caller claims to be a telephone engineer who has called to test your phone and you have not made any complaint about it to British Telecom, refuse admittance or make the caller wait outside while you check your phone – having first shut and locked the door. Adopt exactly the same procedure if a caller claims he is an official sent to check for a gas leak, or to look at the drains. Burglars often adopt one of these guises to obtain entry without force. If they think, once inside, that they can frighten you into taking no action while they look for valuables which they can carry away, they will.

Bogus callers may trick you into letting them go to another part of the house while they ask you to stay where you are to listen for some noise or movement which they specify. While you innocently carry out their instructions they will search for any valuables, or perhaps find the hiding place where you keep that little hoard of money which might be to pay the rent, or the television licence, or may even be your life savings. You will find out that you have been robbed only after they have gone, leaving no trace.

Then there is the man who calls with a tale about how bad your roof looks and then offers to put things right straight away for a very reasonable price. Never have anything to do with such a visitor. This is one of the easiest and most common methods used to trick house-holders into paying for work which is not done because once the man is on your roof you have no way of seeing what he does.

If any caller is persistent, or if there is the slightest attempt to push his or her way in, or if you are in any way suspicious, just tell the stranger that unless he or she leaves at once you will call the police and, if they still persist, do so.

If you carry a shrill personal alarm, do not hesitate to use it if the caller frightens or intimidates you in any way.

If you have a burglar alarm system with a panic button near the door, remember that it is a device which will operate even if the alarm system is switched off, so push it at once and raise the alarm.

If you are thinking about having an alarm system installed, remember that you may first get free advice from the watchdog body for the security industry – the NSCIA – and from your local crime prevention officer.

If you really want to help yourself, your neighbours and the whole local community to beat the burglar and the vandal, remember that there may be a Neighbourhood Watch scheme in your area. Make inquiries and ask if you can join it. If not, you can certainly co-operate with the scheme and, in any case, if you and your neighbours keep an eye on each other's homes and report anything suspicious at once you will be helping to keep every home safe and sound.

Remember, if you see or hear anything suspicious, or think that a crime is being committed, dial 999. If you are very suspicious of someone you see hanging about near your home, or that of a neigh-bour, try to memorise a description of that person such as whether the face is long, or thin, or round; whether it is a clean-shaven man, or whether he has a beard or a moustache; if there are any marks of identification, such as scars or tattoos; the colour of the skin; the height, build, age, and complexion; the colour and length of the hair and whether it is straight, curly or receding; whether the person wears glasses; if you have seen him or her at close quarters, the colour of the eyes; what clothing they are wearing. This is the information the police want to know. If it is possible, write down this information at the time so that you do not forget or make mistakes later.

If you are suspicious of a vehicle, make a note of the make and model, of the colour and registration number and pay particular attention to whether there are any distinguishing marks such as damage or the name of a company.

Finally, keep a note of the telephone number of your local police station near your telephone just in case you need to make a call.

Check list

WHEN YOU ARE GOING OUT

1. Have you closed and locked all doors and windows?

2. If you have a burglar alarm, have you remembered to switch it on?

3. If you are going out in the car, have you remembered to shut and lock the garage doors?

WHEN YOU ARE GOING ON HOLIDAY

1. Have you cancelled delivery of the newspapers?

2. Have you cancelled delivery of the milk?

3. Have you told the police you are going away and given them the name and address of a key holder?

4. Have you remembered to give a spare key to a neighbour or relative who lives nearby, and told them that you have informed the police they have the key?

5. Have you asked the next door neighbour to remove any mail or other literature which might be left sticking out of your letterbox?

6. Have you locked the garden shed?

7. Have you made sure that there are no loose ladders lying around?

8. Have you set an automatic timeswitch in at least one room in the front of the house, plugged it in and made certain it is switched on so that the lights will come on automatically when it is dark?

9. Have you unplugged all electrical appliances which are not going to be in use?

10. Have you made a final check that you have closed and locked all doors and windows?

YOUR PROPERTY

1. Have you taken a photograph of all your valuable possessions such as jewellery, ornaments and antiques?

2. Have you marked all your property with your postcode and house number?

3. Have you made a list of the serial numbers of all such equipment as television sets, hi-fis, videos, computers and cameras?

4. Have you made certain that all your door and window locks are adequate, and renewed any that are old, worn or insecure?

YOUR CAR

1. Have you had some form of anti-theft device fitted to your car?

2. Do you always remember to remove the ignition key when you leave your car?

3. Do you remember to lock the boot and all doors and to shut all windows every time you leave the vehicle?

4. Do you remove all valuables before you leave the car parked anywhere, or lock them in the boot?

Address list

PRODUCT GUIDE

Abloy Locking Devices Ltd, 309–313 West End Lane, London, NW6 1RU.

Adams-Rite (Europe) Ltd, Unit 6, Moreton Industrial Estate, London Road, Swanley, Kent, BR8 8TZ.

Ademco Sontrix Ltd, 11 Cradock Road, Reading, Berkshire, RG2 0JT.

ADT Security Systems, Edgbaston House, 3 Duchess Place, Edgbaston, Birmingham, B16 8NH.

AFA-Minerva, Security House, Grosvenor Road, Twickenham, Middlesex, TW1 4AB.

Aircall Ltd, Radio and Telephone Answering Services, 176 Vauxhall Bridge Road, London, SW1.

Alcan Windows Ltd, Goodman Street, Leeds, West Yorkshire, LS10 1QN.

Alert Systems Ltd, 19–21 Nile Street, London, N1.

Banham Locks and Alarms Ltd, 233–235 Kensington High Street, London, W8 6SF.

Barkway Electronics, The Melbourn Science Park, Melbourn, Royston, Hertfordshire.

Berol Ltd, Oldmedow Road, King's Lynn, Norfolk, PE30 4JR.

Bolton Gate Co Ltd, Turton Street, Bolton, Lancashire, BL1 2SP.

Britannia Security Systems Ltd, Florence Road, Maidstone, Kent.

Burt Boulton Architectural Co, Burts Wharf, Crabtree Manorway, Belvedere, Kent, DA17 6BD.

Car Sounds, 861 High Road, Goodmayes, Ilford, Essex.

Care Trust, Care House, Bigland Street, London, E1 2ND.

Castell Safety International Ltd, Kingsbury Works, Kingsbury Road, London, NW9.

Central Monitoring Services Ltd, 2 Seagrave Road, London, SW6 1RP.

Channel 1 Associates Ltd, 34 Rockingham Road, Uxbridge, Middlesex, UB8 2TZ.

Chubb Alarms Ltd, 42–50 Hersham Road, Walton-on-Thames, Surrey, KT12 1RY.

Chubb Home Protection, 29 Enford Street, London, W1.

Chubb and Sons Lock and Safe Co Ltd, 51 Whitfield Street, London, W1P 6AA.

Churchill Lock and Safe Co, Brymbo Road Industrial Estate, Holditch, Newcastle-under-Lyme, Staffordshire, ST5 9HX.

ComforTec Windows Ltd, Unit 10, British Wharf, London, SE14 5RS.

Copydex Ltd, 1 Torquay Street, Harrow Road, London, W2 5EL.

Crittall Windows Ltd, Manor Works, Braintree, Essex, CM7 6DF.

Data Design Techniques Ltd, 68–70 Tewin Road, Welwyn Garden City, Hertfordshire, AL7 1BD.

Davco Instrumentation and Security Co Ltd, Unit 61, B/2 Farraday Way, Westminster Industrial Estate, London, SE18 5TR.

Davis Security Communications Ltd, Wellsyke Road, Adwick, Doncaster, South Yorkshire.

Eagle International, Precision Centre, Heath Park Drive, Wembley, Middlesex, HA0 1SV.

Etching Transfer Co Ltd, 8 Windlesham Road, Shoreham, West Sussex, BN4 5AE.

First Inertial Systems, Molesey Avenue, West Molesey, Surrey.

Floral Wire Products (Brighouse) Ltd, Caldervale Works, River Street, Brighouse, West Yorkshire, HD6 1JS.

Fullex (Windowcraft) Ltd, Building 57, Third Avenue, Pensnett Trading Estate, Kingswinford, West Midlands, DY6 7PP.

Gamma Electronics, 125 Golders Green Road, London, NW11.

Geemarc Ltd, Lawford House, 1–3 Albert Place, London, N3 1QB.

Gent Ltd, Temple Road, Leicester, LE5 4JF.

Group 4 (Securitas), Farncombe House, Broadway, Worcestershire, WR12 7LJ.

Guardall Ltd, Alexandra Road, Enfield, Middlesex, EN3 7ER.

Guardmark Ltd, Security House, 18 New North Parade, Huddersfield, West Yorkshire, HD1 5JP.

Guiness Security Systems Ltd, 21 Brompton Arcade, London, SW3.

Haley Radio Security Ltd, 8 Gainsborough Road, London, E11 1HT.

Hamber and Whiskin Engineering, Radford Way, Billericay, Essex.
Homesitters Ltd, Moat Farm, Buckland, near Aylesbury, Buckinghamshire, HP22 5HY.
Hoover Home Security, Perivale, Greenford, Middlesex.
Ingersoll Locks Ltd, Forsyth Road, Woking, Surrey, GU21 5RS.
Intercept Alarms Ltd, Unit 4, Central Park Estate, Staines Road, Hounslow, Middlesex, TW4 5DJ.
Kaba Locks Ltd, Woodward Road, Howden Industrial Estate, Tiverton, Devon, EX16 5HW.
Kalami Ltd, Prospect House, The Broadway, Farnham Common, Slough, Berkshire, SL2 3PQ.
Keyco, Arndale House, 19 High Road, Maltby, Rotherham, South Yorkshire.
Krooklok Automotive Division (Dana Ltd), Greenbridge Road, Swindon, Wiltshire.
Lander Alarms Ltd, Sunley House, 46 Jewry Street, Winchester, Hampshire, SO23 8RP.
J. Legge and Co Ltd, Willenhall, West Midlands, WB13 1TD.
Lionweld Ltd, Marsh Road, Middlesbrough, Cleveland, TS1 5JS.
Louvre Lock Co, 118 Stockwell Road, London, SW9 9HR.
Magnet and Southerns Ltd, Royd Ings Avenue, Keighley, West Yorkshire, BD21 4BY.
Minder Products Co, 90 Wallis Road, London, E9 5LN.
Modern Alarms Ltd, 25–26 Hampstead High Street, London, NW3 10A.
Monarch Aluminium Ltd, Kingsditch Lane, Cheltenham, Gloucestershire, GL51 9PB.
Monitron Computer Monitoring Services Ltd, 52 Ebury Street, London, SW1W 0LW.
Motolarm (Capital Car Radio Ltd), 3 Broadwell Parade, Broadhurst Gardens, London, NW6 3BQ.
Network Communications Services Ltd, Michael House, Chase Side, London, N14.
Noise and Security Appliances Ltd, Byron House, Wallingford Road, Uxbridge, Middlesex.
Notecalm Ltd, 4 Goldington Road, Bedford, MK40 3NF.
Panasonic, 300–318 Bath Road, Slough, Berkshire.
Peak Technologies Ltd, Dayston Works, Warwick Road, Boreham Wood, Hertfordshire, WD6 1NA.
Personal Alarms Ltd, Abbotshill, Durley Park, Keynsham, Bristol, Avon, BS18 2AT.
Philips Service, 604 Purley Way, Waddon, Croydon, Surrey, CR9 4DR.
Pifco Ltd, Failsworth, Manchester, M35 0HS.
Pyke International Ltd, Haywood House, 64 High Street, Pinner, Middlesex, HA5 5QA.
Quirefive Ltd, 7 Berners Mews, London, W1P 3DG.
Racal Security Ltd, Lochend Industrial Estate, Newbridge, Midlothian, EH28 8LP, Scotland.
Ramicube Ltd, Unit 442, Walton Summit Centre, Bamber Bridge, Preston, Lancashire, PR5 8AU.
Rothtron Electronics, 23 Havelock Street, Desborough, Northamptonshire.
Safe and Sound Products, PO Box 33, Epping, Essex, CM16 4AN.
Salsbury Locks Ltd, 302 High Street, Croydon, Surrey, CR0 1NG.
Saracen Safes and Security Co Ltd, 67 Nine Mile Ride, Wokingham, Berkshire.
Secure Safes (Coventry) Ltd, Althorpe Street Trading Estate, Leamington Spa, Warwickshire, CV31 2AU.
Securicor-Granley, Auckland House, New Zealand Avenue, Walton-on-Thames, Surrey, KT12 1PL.
Securitas (Group 4 Securitas), Farncombe House, Broadway, Worcestershire.
Security Equipment Installations, Cinderpath, Broadstairs, Kent.
Selmar Burglar Alarm Co Ltd, The Causeway, Maldon, Essex.
Simba Security Systems Ltd, Security House, Occupation Road, London, SE17 3BE.
Songuard Ltd, Mill Mead, Staines, Middlesex, TW18 4UQ.
Status Electronics Ltd, 148 Loughton Way, Buckhurst Hill, Essex, IG9 6AR.
Sterling Guards Ltd, Sterling House, Empress Place, London, SW6 1TT.
Stoughton Lodge (Security) Ltd, Stoughton Lodge, 2 Davies Avenue, Leeds, West Yorkshire, LS8 1JY.
Superswitch Electric Appliances Ltd, 7 Station Trading Estate, Camberley, Surrey, GU17 9AH.
3M (UK), PO Box 1, Bracknell, Berkshire.
Thrust Technology Ltd, 2 Hornsby Square, Basildon, Essex, SS15 6NZ.
Thunder Screw Anchors Ltd, Victoria Way, Burgess Hill, West Sussex.

Time and Data Systems International Ltd, Crestworth House, Sterte Avenue, Poole, Dorset, BH15 2AL.
Trends Security Alarm Systems Ltd, Norfolk House, 116 Western Road, Brighton, East Sussex, BN1 2AB.
Triplex Safety Glass Co Ltd, Eckersall Road, Kings Norton, Birmingham, B8 8SR.
Videoalert (Alert Products Ltd), 32 Hotchkiss Way, Binley Industrial Estate, Coventry, West Midlands, CV3 2RL.
Vigilante (Reliance Security Services) Ltd, Field Street, London, WC1X 9DA.
Vital Communications (UK) Ltd, 91–93 Queens Road, London, SE15 2EZ.
Volumatic Ltd, Taurus House, Kingfield Road, Coventry, West Midlands.
Waso Security Systems, Whiteway Road, Queenborough, Kent, ME11 5EQ.
Yale Security Products Ltd, Wood Street, Willenhall, West Midlands, WV13 1LA.

ORGANISATIONS

Independent Association of Alarm Installers, 2 Command Road, South Gosforth, Newcastle upon Tyne.
National Association of Victim Support Schemes, 169 Clapham Road, London, SW9.
National Supervisory Council for Intruder Alarms, St Ives House, St Ives Road, Maidenhead, Berkshire, SL6 1RD.

PUBLICATIONS

Crime Prevention News, Home Office, Queen Anne's Gate, London, SW1H 9AT.
Home Security and Personal Protection, Security Publications Ltd, PO Box 58, Chislehurst, Kent, BR7 5SQ.

EXHIBITIONS

IFSSEC (International Fire, Security and Safety Exhibition and Conferences Ltd), Cavendish House, 128–134 Cleveland Street, London, W1P 5DN.

Picture Acknowledgements

The author and publishers gratefully acknowledge the help of the following companies who provided the photographs which have been used in this book: *page 6* Abloy Locking Devices Ltd; *page 20* Yale Security Products; *page 22* Banham Locks and Alarms Ltd; *page 24* Yale Security Products; *page 26* Chubb & Sons Lock and Safe Co; *page 28* Chubb & Sons Lock and Safe Co; *page 30* Yale Security Products; *page 31* Chubb & Sons Lock and Safe Co; *page 32* Chubb & Sons Lock and Safe Co; *page 36* Minder Products Co; *page 37* Racal Security Ltd; *page 38* Banham Locks and Alarms Ltd; *page 39* Banham Locks and Alarms Ltd; *page 40* Eagle International; *page 42* Noise and Security Appliances Ltd; *page 43* Notecalm Ltd; *page 44* Intercept Alarms Ltd; *page 45* Kalami Ltd; *page 47* AFA-Minerva; *page 51* Alert Systems Ltd; *page 52* Alert Systems Ltd; *page 54* Haley Radio Security Ltd; *page 55* Davis Security Communications Ltd; *page 59* Barkway Electronics; *page 60* Thrust Technology Ltd; *page 64* Bolton Gate Co Ltd; *page 71* Data Design Techniques Ltd; *page 72* Songuard Ltd; *page 73* Quirefive Ltd; *page 74 (top)* Hamber and Whiskin Engineering; *page 74 (below)* Hoover Home Security; *page 76* The Post Office; *page 78* Guardmark Ltd; *page 79* Superswitch Electric Appliances Ltd; *page 84* Banham Locks and Alarms Ltd; *page 85* Chubb & Sons Lock and Safe Co; *Page 86* Banham Locks and Alarms Ltd; *page 87* Banham Locks and Alarms Ltd; *page 88* Chubb & Sons Lock and Safe Co; *page 89* Banham Locks and Alarms Ltd; *page 93* Simba Security Systems Ltd; *page 104* Pifco Ltd; *page 105* Gent Ltd.

Index